The Electronic Packaging Series

Series Editor: Michael G. Pecht, University of Maryland

Advanced Routing of Electronic Modules
Michael Pecht and Yeun Tsun Wong

Electronic Packaging Materials and Their Properties
Michael Pecht, Rakesh Agarwal, Patrick McCluskey, Terrance Dishongh,
Sirus Javadpour, and Rahul Mahajan

Guidebook for Managing Silicon Chip Reliability
Michael Pecht, Riko Radojcic, and Gopal Rao

High Temperature Electronics
Patrick McCluskey, Thomas Podlesak, and Richard Grzybowski

Influence of Temperature on Microelectronics and System Reliability
Pradeep Lall, Michael Pecht, and Edward Hakim

Long-Term Non-Operating Reliability of Electronic Products
Michael Pecht and Judy Pecht

ELECTRONIC PACKAGING

Materials and Their Properties

Michael G. Pecht
CALCE Electronics Packaging Research Center
University of Maryland, College Park

Rakesh Agarwal
Delco Electronics, Kokoma, Indiana

Patrick McCluskey

Terrance Dishongh

Sirus Javadpour

Rahul Mahajan
CALCE Electronics Packaging Research Center
University of Maryland, College Park

CRC Press
Boca Raton   London   New York   Washington, D.C.

Portions of the text were adapted from R.J. Hanneman, A.D. Kraus, and Michael Pecht, editors, *Physical Architecture of VLSI Systems*, 1994. Adapted by permission of John Wiley & Sons, Inc.

Library of Congress Cataloging-in-Publication Data

Electronic packaging materials and their properties / Michael G. Pecht
    … [et al.].
        p.   cm.-- (Electronic packaging series)
        Includes bibliographical references and index.
        ISBN 0-8493-9625-5 (alk. paper)
        1. Electronic packaging-- Materials. I. Pecht, Michael.
    II. Series.
    TK7870.15.E4222 1998
    621.381′046—dc21   98-34479
                                                                                 CIP

Table of Contents

Preface

The effectiveness with which an electronic system performs its electrical functions, as well as the reliability and cost of the system, are strongly determined not only by the electrical design, but also by the packaging materials. Electronic packaging refers to the packaging of integrated circuit (IC) chips (dies), their interconnections for signal and power transmission and heat dissipation. Packaging is also required for electromagnetic interference (EMI) shielding. In electronic systems, packaging materials may also serve as electrical conductors or insulators, provide structure and form, provide thermal paths and protect the circuits from environmental factors such as moisture, contamination, hostile chemicals and radiation. As the speed and power of electronics increase, the heat dissipation problems and the signal delay caused by the capacitive effect of the dielectric material become even greater issues that need resolving. The solution involves the devising of innovative packaging schemes and the continuing search for more advanced materials.

The level of packaging or packaging architecture is often used to classify materials and the required material characteristics for effective performance over time. The chip, component, printed wiring board, and assembly level packaging are referred to as the zeroth, first, second and third levels of packaging, respectively (the fourth and fifth levels of packaging being the electronic module formation by the integration of the backpanel and power supply with an outer housing and the system formation by integration of electronic modules, e.g., peripherals). In general, each level has unique material properties requirements. The actual applications of materials in electronic packaging include interconnections, printed circuit boards, substrates, encapsulants, interlayer dielectrics, die attach materials, electrical contacts, connectors, thermal interface materials, heat sinks, solders, brazes, lids and housings.

The dependence of material properties on orientation, with respect to material axes, is often indicated by referring to materials as isotropic, orthotropic, or anisotropic. The materials of concern here are mostly isotropic, exhibiting the same behavior along all directions. However, semiconductors and composite materials are often orthotropic and exhibit mutually independent material properties along three mutually perpendicular axes. An orthotropic property of a semiconductor is generally referred to with respect to crystallographic structure. Composite materials are generally referred to with respect to the microstructure of the material architecture.

In this book, materials employed in the zeroth-level packaging are covered in Chapter 2, the first-level packaging materials in Chapter 3, the second-level packaging materials in Chapter 4, and the third-level packaging materials in Chapter 5. Prior to these chapters is an overview of the key electrical, thermal and thermomechanical, mechanical, chemical, and miscellaneous properties and their significance in electronic packaging.

1 PROPERTIES OF ELECTRONIC PACKAGING MATERIALS

1.1 Electrical Properties

Signal processing is critical for the operation of an electronic system, and materials, along with their architectures, play an important role in the propagation of signals, especially for circuits operating at high speeds and at high electrical frequency. The electrical properties of major importance in material selection include the dielectric constant, loss tangent, dielectric strength, volumetric resistivity, surface resistance, and arc resistance. These electrical properties are defined and discussed with standard test methods (refer to Table 1) wherever applicable. Some of these properties exhibit subtle differences and some are known by more than one term.

Dielectric constant, ε. The dielectric constant of an insulating material is the ratio of the measured capacitance with the dielectric material between two electrodes to the capacitance with a vacuum or free space between the electrodes. The dielectric constant is a dimensionless number also referred to as the relative permittivity. Table 2 lists dielectric constants of some electronic. Test method ASTM D150 is used to measure the dielectric constant.

To account for the electrical power loss to an insulating material subject to a sinusoidally time-varying applied potential, a complex number called permittivity is defined as

$$\varepsilon = \varepsilon' - \varepsilon'' \qquad (1)$$

where the imaginary part ε'' is the electrical power loss factor, and the real part, ε', is the dielectric constant of the insulator material.

Most materials possess a dielectric constant that depends on the frequency of the applied electromagnetic field. The dependence on electrical frequency is due to the orientational (defined by reorientation of inherent dipoles in the material), ionic (defined by displacement of ions of molecules), and electronic polarization (defined by the shift in the electronic cloud of atoms) of the material. Figure 1A shows the typical trends of the dielectric constant as a function of frequency for a material with all three types of polarizations.

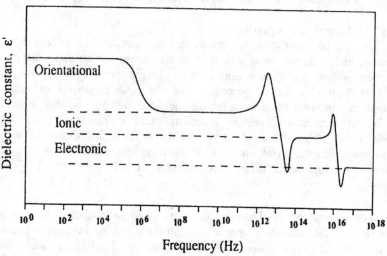

Figure 1A : Diagram of variation in dielectric constant as a function of frequency for a material with orientational, ionic and electronic polarization (Zaky, 1970).

The dielectric constant typically increases with decreased temperature. Because water possesses a rather high dielectric constant (greater than most electronic materials), the dielectric constant of a material increases as the absorbed moisture increases.

The delay of signal propagation through a conductor depends on the dielectric constant of the insulating material. The propagation delay measured in nanoseconds through one foot of conductor wire for a micro-strip is

$$t_{pd} = 1.017\sqrt{0.475\varepsilon' + 0.67} \tag{2}$$

and for a strip-line is

$$t_{pd} = 1.017\sqrt{\varepsilon'} \tag{3}$$

Equations (2) and (3) show that the propagation delay varies directly with the dielectric constant. For example, a reduction of dielectric constant by a factor of 0.5 reduces 30% of the effective conductor length for the signal for a strip-line and slightly less for a micro-strip. To achieve smaller propagation delays in high-speed applications, low dielectric constant materials are sandwiched between signal and ground planes. Because a semiconductor material generally has a higher dielectric constant than the mounting platform material, a signal conductor should leave the surface of an integrated circuit as soon as possible. A high dielectric constant material is desirable between the ground and the power plane to smooth small inductive spikes in power.

Table 1. Standard test specifications for property testing

Property	ASTM Specification
Electrical properties	
Dielectric constant or relative permittivity	D150, D2149, D2520, D5109
Dissipation factor or loss tangent	D150, D2149, D2520
Dielectric strength or breakdown strength	D149, D352
Volumetric resistivity or bulk resistivity	D257, D1829
Surface resistivity	D257, D1829
Insulation resistance	D257, D1829 , MIL-STD-883 METHOD 1003
Arc resistance	D495
Thermal and Thermomechanical properties	
Glass transition temperature	E1363
Deflection temperature	D648
Thermal conductivity	C408
Coefficient of thermal expansion	D696, E228
Mechanical properties	
Elastic modulus	D638
Poisson's ratio	C623
Flexural modulus	D790, C674, F417
Flexural strength	D790, C674, F417
Tensile strength	D638
Creep	D2990
Fatigue endurance	D671
Chemical properties	
Water absorption	D570, MIL-P-13949F (D-24/23)
Solvent uptake	E1209
Resistance to acids and alkalis	D543
Corrosion	C871
Miscellaneous	
Specific gravity	D792, C329
Solderability	-
Toxicity	-

Loss factor, ε". The loss factor is a dimensionless number given by the imaginary part of the complex dielectric permittivity, as defined in Equation(1). The negative sign indicates the loss of electrical power from the conductor. The lost electrical power absorbed per unit volume of the insulator is given by the product of ε'' and the power through the conductor. A low loss factor is desirable for a dielectric material so that the dissipated electric power to the insulator is minimized. This type of consideration is very important for high power circuits operating at high speeds.

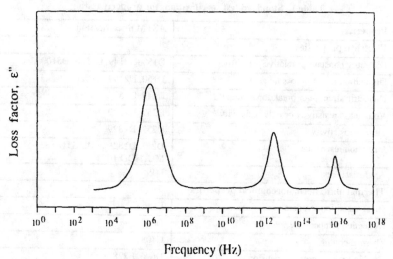

Figure 1B: Diagram of variation in loss factor as a function of frequency for a material with orientational, ionic, and electronic polarization (Zaky 1970).

The loss factor is a function of material conditions such as bulk temperature and absorbed moisture. Variations in the loss factor as a function of electrical frequency depend on the molecular structure of the material. Figure 1B shows the dependence trends of the loss factor on frequency for materials with orientational, ionic and electronic polarization.

Loss tangent or dissipation factor, tanδ.   The loss tangent is a dimensionless number given in terms of the dielectric constant and the loss factor by the relation

$$\tan \delta = \frac{\varepsilon''}{\varepsilon'} \qquad (4)$$

Note that the power dissipation in a dielectric material is proportional to the product of the loss tangent and the dielectric constant. For small loss angles typical of electrical insulation materials, the loss tangent is practically equal to the loss angle expressed in radians. The loss tangent varies not only with the conditions of the dielectric, such as bulk temperature and moisture content, but also with the frequency of the applied electromagnetic field. The loss tangent for several electronic materials is given in Table 2. For most plastics the loss tangent tends to decrease with increasing frequency, but to increase with increasing temperature. Test method ASTM D150 covers the measurement of the loss tangent.

Table 2. Dielectric constant and loss tangent of electronic materials

Material	Dielectric constant @ 25°C and 1 MHz	Loss tangent (x 10^4) @ 25°C and 1MHz
Quartz (Fused silica)	3.8	2
E-glass	6.3	37
S-glass	4.6	15
Nylon	3.3	235
Polyethylene	2.25	1
P.V.C.	3.2 – 3.6	450 – 900
Teflon	2.1	2
Polyimide	3.4 – 3.5	0.0025 –0.01
Polystyrene	2.6	0.7
Epoxy resin	3.5 – 4	300

Power factor. The power factor is the ratio of the power absorbed by the insulator to the total power through the conductor. The power factor is expressed as the cosine of the phase angle q, or the sine of the loss angle d:

$$power \ \ factor = \frac{P}{VI} = \cos q = \sin d \qquad (5)$$

where P is the power absorbed by the dielectric, and I is the current flowing through the capacitor in charging it to a voltage, V. For a small angle, d,

$$\sin \delta \approx \tan \delta \approx \delta \ \ \text{(in radians)} \qquad (6)$$

When the loss tangent is less than 0.1, the power factor differs from the loss tangent by less than 0.5%. Lost electric power is dissipated either as electromagnetic radiation or heat. The magnitude of the energy lost as electromagnetic radiation is typically very low compared with the heat generated. However, measures to shield the electromagnetic radiation are important because radiation can be a hazard to health and causes electrical interference.

Dielectric strength, V_{DS}. The dielectric strength, expressed in megavolts per cm (MV/cm), is defined as the voltage gradient that an insulating material can withstand before an arc forms through the material. The dielectric strength is also known as the breakdown strength. Dielectric materials break down at certain levels of applied voltage gradient for a given set of temperature and humidity conditions. The dielectric strength is inversely proportional to the thickness of the test specimen, that is, thick test pieces have a lower dielectric strength than thin pieces. The higher the dielectric strength, the better is the insulation against applied voltage gradients. Tests are conducted either by increasing the voltage from zero to breakdown in a "short-time test" or in specified increments for a "step-by-step test" [ASTM D149].

Electrical resistivity, r.   Electrical resistivity is the characteristic of a material that resists the passage of electric current. This may be through its bulk or on its surface. It is expressed in ohm-cm. Its two components are volumetric and surface resistivity. Electrical resistivity of some materials is given in Table 3.

Surface resistivity, r_S.   Surface resistivity is the resistance to electrical conduction between electrodes at two opposite edges of a unit square centimeter surface film.   The unit of surface resistivity is ohms (Ω). However, in order to distinguish surface resistivity from resistance, surface resistivity is expressed in ohms per square.   The mechanism of surface conduction is through an adsorbed film on the surface of the material. Surface contaminants, ambient temperature, and humidity affect surface resistance.   A material with higher surface resistance provides better insulation against electrical conduction on its surface. Specifications ASTM D257 and ASTM D1829 describe the test methods for surface resistivity measurements.

Volumetric resistivity, r_V. Volumetric resistivity, expressed in $M\Omega$-cm is the electrical resistance between opposite faces of a unit cube when current flow is confined to the specimen volume.

Table 3. **Electrical resistivity of various packaging materials**

Material	Electrical Resistivity (Ω-cm)
Aluminum Film	3.225×10^{-6}
Chromium Film	$0.125 - 0.274 \times 10^{-6}$
Copper Film	2.066×10^{-6}
Nickel Film	8.44×10^{-6}
Titanium Film	0.433×10^{-6}
Copper	1.725×10^{-6}
Diamond	9.65×10^{7}
Molybdenum	5.52×10^{-6}
Aluminum	2.733×10^{-6}
Tungsten	5.44×10^{-6}
Kovar (Fe + 29 Ni + 17 Co)	0.495×10^{-6}
Pb + Sn (95/ 5)	0.214×10^{-6}
Silicon Carbide	2.453×10^{9}
Silicon	2.93×10^{5}
Gallium Arsenide	3.0×10^{8}
Gold	2.27×10^{-6}

Volumetric resistivity is a bulk material property that varies with temperature, moisture content, applied voltage and the time duration of the applied voltage. Volumetric resistivity and resistance are related by the expression

$$\rho_v = \frac{RA}{l} \tag{7}$$

where r_v is the volumetric resistivity, R is the electrical resistance, A is the effective cross-section area, and l is the effective length of the test piece. Resistance is expressed in ohms and is dependent on the geometry of the conductor. In general, a material with higher resistivity is a better insulator. Specifications ASTM D 257 and ASTM D 1829 describe the test methods for volumetric resistivity measurements.

Insulation resistance, IR. Insulation resistance is more a measure of the quality of insulation provided by a packaging architecture than a real material property. Although expressed in terms of leakage current, with units of nano amperes (nA), IR measures the resistance offered by an insulating material to a direct impressed voltage across the lead and case of the component part. It is useful for comparative evaluations of materials for given test conditions. Both volumetric and surface resistances contribute to the leakage current. Low insulation resistance indicates inadequate insulation of the conductor paths. The tests are performed according to Test Method 1003 of MIL-STD 883C.

Arc resistance, Dt$_{arc}$. The arc resistance of a material is defined as the elapsed time before the action of an arc at the surface of the material forms a conducting path. The arc resistance, expressed in seconds, is indicative of the ability of the material to resist the formation of an arc. Materials that form a permanent carbonized path due to the action of an arc are said to "track". Arc resistance is affected by surface conditions such as cleanliness and dryness. A high value of arc resistance is desirable to avoid electric breakdown along the insulation surface through arcing. The test method for arc resistance is described in ASTM D 495.

1.2   Thermal and Thermomechanical Properties

An electronic material experiences a range of steady-state temperatures, temperature gradients, rates of temperature change, temperature cycles, and thermal shocks through manufacturing, storage, and operation. Thermal properties of electronic materials that are significant in enduring such life cycle profiles include thermal conductivity, deflection temperature, glass transition temperature (typical for polymeric materials), and the coefficient of thermal expansion.

Deflection temperature, T_d. Deflection temperature is the temperature at which a specified deflection occurs in a specimen under a selected load and loading method. Test method ASTM D 648 specifies the loading for deflection temperature measurements. The temperature is indicative of the mechanical load-carrying capability of a material. The deflection temperature of polymeric materials and composites may approximate their glass transition temperature, especially for thermoplastics. For a thermoset polymer, the degree of cure can lead to significant differences between deflection temperature and glass transition temperature. In the case of reinforced laminates, the relief of residual stresses can further complicate the correlation.

Thermal conductivity, k. Heat conduction is described by Fourier's Law. For one-dimensional heat conduction through a plane wall with a temperature distribution $T(x)$, the heat equation is expressed as

$$q_x = -k\frac{dT}{dx} \tag{8}$$

where q_x (W/m^2) is the heat transfer rate per unit area perpendicular to the direction of the heat flow, and dT/dx (°C/m) is the temperature gradient in the x-direction. The proportionality constant k is a heat transport property known as thermal conductivity, expressed in Watts per meter per degree centigrade (W/m-°C). The minus sign in eq. (8) is required to assure positive heat flow in the direction of a falling or negative temperature gradient.

Table 4. Comparison of thermal conductivity and resistivity of various polymers, ceramics and metals

Material	Thermal conductivity, k (W/m-°C)	Volumetric resistivity, (µΩ-cm)
Diamond	2000	10^{22}
Copper	380	0.34
Berylia	286	10^{20}
Aluminum	210	0.52
Aluminum nitride	200	10^{19}
Molybdenum	130	1
Lead	36	4.4
Alumina	30	10^{20}
Polyurethane	3.3	10^{16}
Epoxy glass	0.3	10^{21}
Polyimide	0.2	10^{20}
Acrylic coating	0.2	10^{21}
Acrylic	0.2	10^{21}

Table 5. Typical glass transition temperature, T_g, values of common resin materials

Resin system	Glass transition temperature, T_g (°C)
Epoxy type FR-4	125 – 135
Polyfunctional epoxy	140 – 150
Epoxy type FR-5	140 – 160
High-temperature epoxy	170 – 180
Bismaleimide triazine epoxy	180 – 190
Cyanate esters	240 – 250
Polyimide	240 – 260
Modified polyimides	260 – 300
Acrylic	40 – 55
PTFE (melting point)	327

Thermal conductivity is often a function of temperature. Test methods ASTM D 792 and ASTM C 408 cover thermal conductivity measurements. Polymers are generally poor conductors and exhibit very low thermal conductivity, often an order or two of magnitude lower than metals. Ceramics exhibit better thermal conductivity than polymers. Table 4 compares the thermal conductivity of polymers, ceramics, and metals. Generally, materials with high electrical resistance possess low thermal conductivity (refer to Table 4). Beryllia (BeO) is an exception to this rule. It possesses good electrically insulating properties and thermal conductivity compared to metals.

The effective thermal conductivity of some materials can be improved by adding fibers or fillers of higher thermal conductivity. An example material with filler enhanced thermal conductivity is silver-filled epoxy.

Glass transition temperature, T_g. The glass transition temperature is a material property of polymers that is generally not exhibited by metals or ceramics. The glass transition temperature is the temperature at which a material changes from a hard, brittle, "glass-like" form to a softer, rubber-like consistency. The change in state occurs over a range of temperature for amorphous polymers. Crystalline polymers such as polytetrafluroethelene (PTFE) exhibit a unique melting point rather than passing through stages of decreasing viscosity with increased temperature. Typical T_gs for common resin materials are listed in Table 5. The test method for glass transition temperature is covered in ASTM E1363.

Electrical, thermomechanical, and mechanical properties of a material dramatically change, often beyond the glass transition temperature. The change is gradual in thermoset polymers relative to thermoplastic polymers because of highly cross-linked molecular chains. The changes in thermoplastics are regained upon cooling below T_g, whereas the changes are irreversible in thermosets because the chemical decomposition that occurs beyond T_g destroys cross-linking among molecular chains.

Coefficient of linear thermal expansion, CTE. CTE is the change in linear dimension per unit length per degree change of the bulk material temperature. It is generally expressed in parts per million-degree centigrade (ppm-°C). The CTE is a manifestation of the volumetric dilation or contraction of a material in a particular direction as a function of the change in the material's bulk temperature. CTE values are normally reported at 55 °C, unless otherwise specified, in accordance with standard test methods ASTM D 696 or ASTM E 228.

The CTE is often a positive quantity, although it can be negative or zero. The CTE can change appreciably with a change in temperature. Polymers exhibit almost an order of magnitude higher CTE than most metals, ceramics, and glasses. Above the glass transition temperature, the CTE of polymers further increases by a factor of 3 to 5.

In the case of polymeric laminates used as mounting platforms for electronic components, it is of considerable interest to develop polymers with glass transition temperatures beyond the temperatures used in manufacturing processes such as soldering. Organic fibers, such as aramid and graphite, are transversely isotropic, exhibiting an order of magnitude higher CTE in the radial direction compared to the axial.

Table 6. Heat capacity of various packaging materials

Material	Heat Capacity (J/g-K)
Aluminum	0.8940
Copper	0.3850
Diamond	0.5180
Gold	0.1284
Molybdenum	0.2502
Tungsten	0.1322
Kovar	0.4320
Pb + Sn (95/ 5)	0.1340
Alumina, Al$_2$O$_3$	0.7790
Aluminum Nitride	0.7450
Mullite / Cordierite Ceramic Composite	0.6900
Silicon	0.7120
Silica, SiO$_2$	0.7440
Silicon Carbide	0.6700
DuPont Kevlar 49	1.450
Polytetrafluoroethylene (PTFE)	1.0283
Gallium Arsenide	0.3220
Water	4.1830

In a structure with hybrid materials, endurance to thermal cycles is dependent on matching CTEs. The improved endurance results from a smaller amplitude of internal stresses because of a smaller dimensional mismatch change caused by the temperature cycling.

Heat capacity, H_c. By definition, the heat capacity of a material is the amount of heat required to raise the temperature of its unit mass by one degree Kelvin. This is an important parameter when evaluating the suitability of a material for a particular packaging application because it determines the quantity of heat that can be dissipated by a specific material. This is more important for power applications where a large amount of heat is required to be dissipated from the die. The typical heat capacities of several packaging materials are given in Table 6.

1.3   Mechanical Properties
The mechanical properties affect the material's ability to sustain loads due to vibrations, shock, and thermomechanical stresses during manufacture, assembly, storage, and operation. Key properties that are of importance for electronic packaging applications include the modulus of elasticity, tensile strength, Poisson's ratio, flexural modulus, fracture toughness, creep resistance, and fatigue strength. However, other properties may be important for special applications.

Modulus of elasticity or Young's modulus, E. When a straight bar is subjected to a relatively small tensile or compressive load, the bar elongates or contracts. The elongation per unit length of the bar is called strain (ε) and the load per unit cross-sectional area is called stress (σ). Let's take a look at the diagram of stress, σ, versus strain, ε, for a typical ductile material.

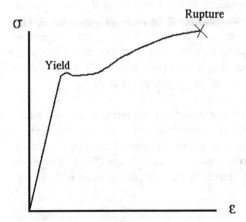

For most materials at small elongations, the stress is proportional to the strain (Hooke's law) before the applied stress achieves the yield stress of the material, such that

$$\sigma = E\varepsilon \qquad (9)$$

where E is the proportionality constant known as modulus of elasticity, or Young's modulus. Because strain is a dimensionless quantity, the units of the modulus of elasticity are the same as stress, often expressed in pascals (Pa). A higher modulus of elasticity indicates a stiffer material. The modulus of elasticity is typically a function of temperature and strain rate. Standard test methods are covered in ASTM D638 and ASTM E111.

Tensile, σ_{ut}, compressive, σ_{uc}, and ultimate strength, σ_u. The tensile or ultimate strength is the maximum stress reached in a stress–strain plot for a standard test, such as that for uniaxial tension. The analogous definition applies to compressive strength for a test in compression. A high value of ultimate strength indicates the ability of the material to carry the stress before rupture.

Because the ultimate strength is easy to determine and quite reproducible, it is useful for the purposes of specification and for quality control of a product. Test method ASTM D638 covers the determination of the ultimate strength.

Flexural modulus (FM). The flexural modulus is the elastic modulus of a material determined from a bending test. Usually a three-point bending test is performed and the flexural modulus is computed as

$$FM = \frac{Pl^3}{4\delta bh^3} \qquad (10)$$

where P is the applied load in MPa, l is the beam length between simple supports in mm, b is the beam width in mm, h is the beam thickness in mm, and d is the beam deflection at the mid point of its length in mm. A high value of flexural modulus indicates a stiff material in flexure or bending, which implies reduced out-of-plane deflection of plates or laminates made from such material under vibrational or transverse loads. With an increase in temperature, the flexural moduli of materials typically decrease.

Poisson's ratio, v. Poisson's ratio is a dimensionless number, defined as the negative of the ratio of transverse to longitudinal strain for uniaxial loading. For isotropic materials, Poisson's ratio ranges between -1 and 0.5. However, for composite materials it can go beyond this range (Jones 1975). Polymers, including adhesives and epoxies, generally possess higher value of Poisson's ratio than ceramics and glasses.

Table 7. Poisson's ratio of some packaging materials

Material	Poisson's Ratio
Aluminum	0.3496
Copper	0.3435
Gold	0.4233
Chromium Film	0.2024
Nickel Film	0.2970
Titanium Film	0.3212
Kovar	0.3360
Alumina, Al(2)O(3)	0.2500
Silicon Nitride	0.2840
Silicon Carbide	0.1560
DuPont Kevlar 49	0.3600
Gallium Arsenide	0.3118
Silicon	0.2782

A mismatch in Poisson's ratio of joined materials can lead to biaxial loading. Test methods ASTM C623 and ASTM E132 cover the measurement of Poisson's ratio. Table 7 gives the value of the Poisson's ratio of some packaging materials.

The state of stress in a bending (or flexural) test specimen is highly non-uniform relative to a uniaxial tension or compression test and involves tensile, compressive and shear stresses. The intensity of these stresses varies through the cross-section of the specimen. The stress state in non-homogeneous materials such as woven fabric composites is further complicated. Tests for flexural modulus are covered in ASTM D790.

Fracture toughness, K_{Ic}. A crack in a solid can be loaded in three different modes – namely, opening mode (mode I), sliding mode (mode II) and tearing mode (mode III). The superposition of the three modes describes the general case of loading. Mode I is the primary cause of failure. The elastic stress field around a crack in a mode I loading can be expressed as (Broek 1986)

$$\sigma_{ij}(r,\theta) = \frac{K_I}{\sqrt{2\pi r}} f_{ij}(\theta) \qquad (11)$$

where s_{ij} denotes the matrix of stress in the solid and f_{ij} are known functions of the angle q from the crack plane. The factor K_I is the stress intensity factor, where the subscript I indicates mode I loading. Fracture toughness, denoted K_{Ic} and expressed in MPa-m$^{\frac{1}{2}}$, is the critical value of stress intensity factor at which crack growth initiates in a material with an inherent crack. Because the thickness of the test specimen plays an important role in the response of the crack, the K_{Ic} value obtained from an experiment is applicable only to a given thickness. The effect is negligible for brittle materials, and K_{Ic} is considered a material property. Materials with a low fracture toughness K_{Ic}, can sustain a smaller crack for the same stress applied at the far field; otherwise, a thicker section is required to ensure against fracture. K_{Ic} often decreases with decreased temperature and increased strain rate. The test procedures for K_{Ic} testing are covered in test methods ASTM E 399 and ASTM B646.

Creep resistance, σ_{cr}, creep strength, σ_{cs}, and rupture strength, σ_{cu}. Many materials continue to deform over time when they are subjected to an applied load (Dieter 1986). Such materials are called viscoelastic. Creep resistance is the ability of a material to resist dimensional change with time under the influence of the applied load. Creep is the time-dependent deformation of a material under constant mechanical stress and temperature. When reporting creep behavior of a material, it is customary to report creep strength and rupture strength. Creep strength is defined as the stress at a given temperature that produces a steady state creep rate of a fixed amount. The creep rupture strength is the stress at a given temperature to produce a life-to-rupture of a certain duration.

In metals, creep is generally negligible at room temperature. Creep can typically not be neglected for thermoplastic polymers and solders. Standard test methods for tensile, compressive, and flexural creep are covered in test method ASTM D 2990. Significant stress relaxation can occur due to creep strains. Creep often needs to be accounted for when estimating fatigue life.

The S–N curve, fatigue strength at N cycles, S_N, and fatigue endurance, S_e. The S–N diagram is a plot of the number of cycles to failure versus the applied stress. The stress may be given as a maximum stress, S_{max}, minimum stress, S_{min}, stress range, S_r, or alternating stress (one half of the range of a cycle), S_a. The number of cycles is usually a mean or median of fatigue life values of a large number of test samples. A common way to establish the S–N curve is to plot alternating stress vs. log N determined from R.R. Moore tests (Shigley 1986).

Fatigue strength is the value of stress for failure at N cycles as determined from the S-N curve. In the case of steels, there is a fatigue strength level below which the material can practically sustain unlimited cycles, known as the fatigue endurance or endurance limit. Nonferrous metals do not exhibit an endurance limit and their fatigue strength is generally reported as corresponding to 10^8 cycles of stress reversal. Plastics generally have a continually decreasing fatigue strength as the number of fatigue cycles increases.

Table 8. **Hardness values of common packaging materials**

Material	Hardness (GPa)
Copper Film	153.3 KP
Kovar (Fe + 29 Ni + 17 Co)	68.0 RW
NEMA FR-4 Glass Fiber/Epoxy Composite	110.0 RW
Aluminum Film	0.587 VR
Diamond Film	72.8 VR
Tungsten Film	19.885 VR
Alumina, Al_2O_3	19.81 VR
Aluminum Nitride	12.20 VR
Silicon Nitride	17.46 VR

KP: Knoop Hardness; RW: Rockwell Hardness; VR: Vickers Hardness

A host of variables affect fatigue life, including temperature, corrosion, mean stress, and residual stresses.   A higher fatigue strength corresponding to a given number of cycles, N, indicates the ability of a material to withstand the formation of a crack or cracks or complete fracture under conditions that produce fluctuating stresses and strains. Test methods ASTM D 671 and ASTM E 1150 cover fatigue. Thermal fatigue occurs when the cyclic loading is due to thermal cycling.

Hardness, H. Hardness is found from measurements of indentation depth caused by pressing a specific tool or indentor into the material surface with a specified force. It is an important parameter when determining the suitability of a material for a particular heavy duty application or when impact resistance is important.   The popularly known methods of measuring hardness are Vickers, Brinells, Knoop, Rockwell and Shore. The hardness values of some electronic packaging materials are given in Table 8.

1.4   Chemical Properties
Chemical properties of electronic materials are important because of the need for electronic materials to survive manufacturing, storage, handling, and operating environments. The chemical properties of significance are water absorption, flammability, and corrosion resistance.

Water (moisture) absorption, Dm. Water absorption is defined as the increase in weight of a material after immersion in water for a specified duration at a specified temperature. The gain in weight of the material is expressed as a percentage of its dry weight. Water absorption is dependent upon the chemical nature of the material as well as the presence of voids. The electrical properties of electronic materials often change as a result of water absorption, and swelling and other dimensional instabilities can occur. Absorbed water in molding material provides the electrolytic medium for potential corrosion of an embedded metal. Test method ASTM D 570 covers the test procedure for determining water absorption.

Flammability (UL 94 classification) and oxygen index (OI). Flammability of a material is specified as a flame retardant grade or as an oxygen index. Flammability is the property of a material whereby flaming combustion is prevented, terminated, or inhibited following application of a flaming or non-flaming source of ignition, with or without subsequent removal of the source.

Underwriters Laboratories material flammability standards (UL 94) assign a flame retardant grade to a material based on its burning rate. The various grades are HB, V-0, V-1, V-2 and 5V, where HB indicates the highest and 5V the lowest burning rate. The numerical value of flammability is given by the oxygen index. The oxygen index is the percent oxygen in an oxygen–nitrogen mixture that will just sustain the combustion of a material. The oxygen index is obtained in compliance with test method ASTM D 2863. Decreasing flaming combustion is indicated by a higher oxygen index.

Polymers are unique in flammability. Polymers such as polyamide-imide, polyetherimide, polysulfone, and polyphenylene sulfide are inherently resistant to flame and require no treatment. Polyamides, polycarbonate, and polysulfone are less resistant. The flammability of polymers can be reduced by compounding with halogenated compounds, phosphate esters, and antimony oxide.

Corrosion resistance. is the ability of a material to resist deterioration due to chemical or electrochemical attack (Davis et al. 1987). Corrosion occurs in metals largely due to the formation of an electrochemical cell and, in non-metals, through chemical attack.

An electrochemical cell is composed of an anode, a cathode, and the medium between these two electrodes, which is referred to as an electrolyte. An electrical potential difference between the anode and the cathode drives the electrochemical reaction, resulting in dissolution of the anode in the form of ions into the electrolyte solution and a continuous flow of electrons between the anode and the cathode. The galvanic series shown in Table 9 is useful in accessing the corrosion resistance of a metal when forming galvanic cells with other metals. The standard oxidation potential (relative to a hydrogen electrode) listed in the table gives the galvanic cell potential of a metal with another metal. The smaller the galvanic potential, the more corrosion resistant is the metal. Corrosion of non-metals leads to formation of more stable compounds and can involve loss in weight, swelling, decomposition, depolymerization, and degradation of physical properties.

Table 9. The galvanic series (Pecht 1991)

Electrode reaction	Standard oxidation potential E (V) @ 25°C
$Li = Li^+ + e-$	3.05
$K = K^+ + e-$	2.93
$Ca = Ca^{++} + 2e-$	2.87
$Na = Na^+ + e-$	2.71
$Mg = Mg^{++} + 2e-$	2.37
$Be = Be^{++} + 2e-$	1.85
$U = U^{+3} + 3e-$	1.80
$Hf = Hf^{+4} + 4e-$	1.70
$Al = Al^{+3} + 3e-$	1.66
$Ti = Ti^{++} + 2e-$	1.63
$Zr = Zr^{+4} + 4e-$	1.53
$Mn = Mn^{++} + 2e-$	1.18
$Nb = Nb^{+3} + 3e-$	1.1
$Zn = Zn^{++} + 2e-$	0.763
$Cr = Cr^{+3} + 3e-$	0.74
$Ga = Ga^{+3} + 3e-$	0.53
$Fe = Fe^{++} + 2e-$	0.440
$Cd = Cd^{++} + 2e-$	0.403
$In = In^{+3} + 3e-$	0.342
$Tl = Tl^+ + e-$	0.336
$Co = Co^{++} + 2e-$	0.277
$Ni = Ni^{++} + 2e-$	0.250
$Mo = Mo^{+3} + 3e-$	0.2
$Sn = Sn^{++} + 2e-$	0.136
$Pb = Pb^{++} + 2e-$	0.126
$H_2 = 2H^+ + 2e-$	0.000
$Cu = Cu^{++} + 2e-$	-0.337
$Cu = Cu^+ + e-$	-0.521
$2Hg = Hg_2^{++} + 2e-$	-0.789
$Ag = Ag^+ + e-$	-0.800
$Hg = Hg^{++} + 2e-$	-0.854
$Pd = Pd^{++} + 2e-$	-0.987
$Pt = Pt^{++} + 2e-$	-1.2
$Au = Au^{+3} + 3e-$	-1.50

1.5   Miscellaneous Properties

Specific gravity, SG; and density, r. Specific gravity and density are often used interchangeably. Specific gravity, a dimensionless number, is the ratio of the mass of a given volume of material at 23°C to the mass of an equivalent volume of water at the same temperature. Density, expressed in grams per cubic centimeter (g/cm^3), is the weight per unit volume of a material at 23°C. The specific gravity and the density of a material differ slightly in numerical value. The relationship between the two at 23 °C is

$$\rho = 0.99756 \ (SG) \tag{12}$$

Specific gravity or density is of special significance in applications when the weight of the structure is required to be low. Test method ASTM D 792 covers the test procedures for obtaining specific gravity of a material.

Toxicity, m_t. Toxicity, expressed in micrograms per cubic meter of air, is the concentration of airborne material which, if inhaled for extended periods of time, is hazardous to human health. For example, the U.S. Government recommends an in-plant maximum of 25 micrograms of beryllia (BeO) per cubic meter of air. Airborne particulates are mostly generated during certain machining operations, such as grinding and laser cutting.

2   ZEROTH-LEVEL PACKAGING MATERIALS

In this section, the discussion centers on the materials and key properties for zeroth-level packaging. The focus is on semiconductor die materials, die attach materials and substrates. These elements make up what is commonly called the die attach assembly.

The electrical properties of a semiconductor determine the functional performance of the device. However, from the viewpoint of electronic packaging engineering, the electrical properties do not affect the design of the die attach assembly. The thermal and mechanical properties of the semiconductor are critical for die attach design. These properties are typically direction-dependent due to the anisotropic nature of the semiconductor crystal.

The material properties involved in the determination of the stresses in the die attach assembly include the modulus of elasticity, Poisson's ratio, and the coefficient of thermal expansion. The key material property involved in the thermal performance of the package is thermal conductivity. The thermomechanical material properties of the semiconductor are fracture toughness, fatigue constants, and flexural strength (modulus of rupture).

2.1   Semiconductors

The digital information age requires the use of different types of semiconductor materials. Materials that have found widespread use in the fabrication of VLSI devices are discussed in this section. The electrical properties of semiconductor materials center around resistivity and dielectric constant. The practical significance of semiconductors lies not so much in their bulk resistivity, but in the fact that the resistivity can be varied by the introduction of atomic amounts of impurities into the crystal structure. By proper material doping, the movement of electrons can be precisely controlled and devices having various functions such as rectification, switching, detection, and modulation can be fabricated. Some electrical, mechanical and thermal properties of semiconductor and die materials are listed in Tables 10, 11 and 12.

Silicon is the most extensively used semiconductor material because, typical of elemental semiconductors, all atoms of the crystal lattice in silicon are of one kind. Gallium arsenide (GaAs) is a compound semiconductor whose crystalline structure consists of gallium atoms alternating with arsenic atoms. Compared with silicon, GaAs has a higher low field mobility, a negative resistance region, a semi-insulating format that results in circuits of monolithic format and the ability to be combined with other III-V compounds in epitaxial growth. However, GaAs has a low fracture toughness to resist the initiation and propagation of cracks during processing, manufacturing and field service operation.

Silicon also forms an extremely stable, high dielectric-strength oxide that is used as an insulator in solid-state devices. This is one of the key attributes of silicon that has led to the development of high-density integrated circuits. Compound semiconductors do not possess stable oxides with

Table 10. Electrical properties of semiconductor and die materials
(Pecht 1991, Lide 1990)

Material	Resistivity (Ω-m)	Dielectric constant @ high frequency
Si	640	11.9
Ge	8.9	16.0
Te	20	-
Diamond	2.7×10^{14}	5.7
SiC	-	10.0
InP	-	12.4
InAs	-	14.6
InSb	-	17.7
AlAs	-	8.5 – 10.9
AlP	-	11.6
AlSb	-	11.0 – 14.4
GaAs	4000	13.1
GaP	-	11.1
GaSb	-	15.7
CdS	-	5.4
CdSe	-	10.0
CdTe	-	7.2 – 10.2
ZnS	2.7×10^{-3} – 1.2×10^{4}	5.2 – 8.9
ZnSe	-	5.9 – 9.2
ZnTe	-	9.0 – 10.4
PbS	6.8×10^{-6} – 0.09	17.0
PbSe	2×10^{-4} – 20.0	24.0 – 161
PbTe	-	30.0 – 360

high dielectric strengths.   Silicon's superior thermal conductivity and mechanical strength also make it a better substrate material. Silicon is also a useful semiconductor when deposited in different phases; amorphous silicon is used in solar cells, CCDs, and ICs; microcrystalline silicon in flat-panel displays; polycrystal silicon in solar cells and ICs; and single crystal silicon is utilized in IC epilayers.

Hydrogenated amorphous silicon (a-Si-H) is an important technological semiconductor thin-film material with applications in large area semiconductors as well as "micro-electronic" devices including photovoltaic cells, liquid crystal displays (LCD), optical scanners, xerographic drums, thin-film transistors, and radiation imaging.

Gallium arsenide (GaAs) is a group of III-V compound semiconductors whose crystalline structure consists of gallium atoms alternating with arsenic atoms (zincblend structure). It has a metal-like appearance with relatively high electron mobility and a large energy bandgap, which make it an ideal material for high-frequency, high-temperature, and radiant-resistance devices.   GaAs is a so-called "direct-bandgap semiconductor," since the minimum of the conduction band and the maximum of the valance band are both at k = 0. Electronic transition between the two bands requires a change in energy only and not in momentum. Electrons can make this transition by absorbing or emitting a photon with a suitable energy. This makes GaAs an excellent candidate for producing a large variety of optical devices such as LEDs, lasers, detectors and so on. GaAs has the ability to be combined with other III-V compounds in epitaxial growth.

$Al_xGa_{1-x}As$ is a ternary-compound semiconductor material. The overwhelming majority of the laser diodes commercially produced until now are based on AlGaAs semiconductor materials. The electronic properties of $Al_xGa_{1-x}As$ depend on its composition. It is a direct bandgap for $0 < x < 0.4$, and indirect for $x > 0.4$. Due to the large energy gap of AlGaAs, it is a very good material for confinement of carriers in GaAs. This is the basis for many quantum well- and two-dimension based devices.

$In_y(Al_xGa_{1-x})_{1-y}P$ quaternary systems are III-V compound semiconductors. The quaternary alloys are lattice-matched to GaAs (at 300 K) for materials in which $y = 0.48$ and $0 < x < 1$. The alloys lattice-matched to GaAs are of the greatest technological interest. The most important applications of these materials are for visible light-emitting diodes (LEDs) and injection lasers, although interest in InGaP/GaAs heterojunction bipolar transistors (HBTs) is increasing.

Germanium (Ge) is another very important semiconductor material. Zone-refining techniques have led to the production of crystalline germanium for semiconductor use with an impurity of only one part in 10^{10}. Ge has a diamond–crystal structure with an indirect bandgap and exhibits a relatively high hole mobility, leading to use in device applications such as high-speed infrared detectors. In semiconductor device technology, Ge and Si, together with GaAs, occupy prominent places.

Ge/Si(100) has been considered a potential substrate for the growth of GaAs due to a good lattice-constant match (within 0.1%) between Ge and GaAs. GaAs substrates are ideal for the growth of bulk Ge devices due to their high resistivity, wide availability, large size, and transparency at the Ge bandgap wavelength.

Si-Ge alloys have been increasingly important for practical applications, including high-speed transistors, long-wavelength infrared detectors, and infrared emitters. There is a 4% difference in the lattice constants of Si and Ge; if one is grown on the other, the layer is strained and must be grown below the critical thickness. This strain may be used to vary the bandgap energy, band discontinuities and effective masses and to split the valley degeneracy and numerous other properties of the material. SiGe material has substantially higher mobility than Si material.

The major advantage of SiGe is that it is compatible with CMOS and hence devices may be designed to be fabricated on a Si chip alongside CMOS and bipolar devices. Hence SiGe devices can have substantially

Table 11. Mechanical properties of semiconductor and die materials
(Pecht 1991, Lide 1990)

Material	Flexural strength (MPa)	Modulus of elasticity (GPa)	Poisson's ratio	Density (kg/m^3)	Knoop hardness (kg/mm^2)
Si	62	109 - 190	0.28	2330	850 – 1150
Ge	-	103	0.28	5340	750 – 780
Te	-	41.3	-	6230	-
Diamond	71.4	-	-	3510	7700
SiC	69	655	0.19	3210	2600
InP	-	-	-	4787	420 – 535
InSb	-	-	-	5775	220 – 225
AlAs	-	-	-	3810	510
AlP	-	-	-	2620	430 – 560
AlSb	-	-	-	4218	360 – 408
GaAs	150	84.95	0.31	5316	535 – 765
GaP	-	-	-	4130	950 – 964
GaSb	-	-	-	5619	450
CdS	-	-	-	4135	55
CdSe	-	-	-	5760	-
CdTe	-	-	-	6200	61
ZnS	-	-	-	4080	180
ZnSe	-	-	-	5420	138
ZnTe	-	-	-	6340	92
PbS	-	-	-	7550	-
PbSe	-	-	-	8120	-
PbTe	-	-	-	8160	-

faster performance than conventional Si transistors while still being produced on Si production lines. As the cost of production lines increases, the line widths shrink; SiGe may be able to provide some solutions.

Wide-bandgap semiconductors such as SiC, III-V nitrides, and related semiconductors are currently attracting increasing attention due to their interesting physical properties, different from conventional semiconductors (like Si and GaAs). Rapid improvement of the material quality and of the knowledge of physical properties is now generating development in high-power, high-temperature, high-frequency electronics and blue light emitters. III-Nitride-based semiconductor compounds, including binary InN, GaN, AlN, ternary $In_xGa_{1-x}N$, $Al_xGa_{1-x}N$ and quaternary InGaAlN, are the most promising material systems for future visible, ultraviolet photodetectors, light-emitting diodes (LEDs), and high-temperature electronics. Currently, available Si and SiC solid-state UV photodetectors suffer from weaknesses such as low quantum efficiency (indirect bandgap), poor solar blindness and fragility in harsh environments.

Tellurides or tellurium compounds, including CdTe, ZnTe, CdZnTe, and CdHgTe, are direct-bandgap II-VI semiconductor materials that are useful in

Table 12. Thermal properties of semiconductor and die materials (Pecht 1991, Lide 1990)

Material	Thermal conductivity (W/m-°C)	CTE (ppm/°C)	Heat capacity (J/kg-°C)	Melting point (°C)
Si	124–148	2.3–4.7	702–712	1685
Ge	64	5.7–6.1	322–335	958–1231
Te	2.0–3.4	16.8	197	723
Diamond	2000–2300	1.0–1.2	471–509	>3823
SiC	283	4.5–4.9	670	3070
InP	67–80	4.5	-	1344
InAs	29	4.7–5.3	268	1216
InSb	16	4.7–5.0	144	808
AlAs	84	3.5	-	>1873
AlP	92	-	-	>1773
AlSb	46–60	4.2	-	1323
GaAs	44–58	5.4–5.72	322	1511
GaP	75–79	5.3	-	1738
GaSb	27–33	6.1–6.9	-	985
CdS	40.1	-	320	1750
CdSe	31.6	-	-	1512
CdTe	6	-	-	1365
ZnS	25–46	-	-	2122
ZnSe	14.0	-	-	1790
ZnTe	10.8	6.6	-	1511
PbS	2.3	-	-	1386
PbSe	1.7	-	-	1338
PbTe	2.3	-	-	1190

a wide variety of electronic and optoelectronic devices ranging from near ultraviolet to far infrared detectors.

Diamond semiconductors, because of their unique physical, optical, and electronic properties, have been proposed for a variety of high-temperature and high-power device applications.

2.2   Attachment Materials
The bond strength of the attachment material is an essential property to ensure that the die and the substrate stay in place despite the stresses imposed during manufacture, storage, and operation. The die must not detach from the substrate during power or temperature cycling or exposure to extreme temperatures. The key material properties for the attachment material are tensile strength, shear strength, and fatigue endurance. For brittle attachment materials, such as glasses, the fracture toughness characterizes the material resistance to fracture. When the attachment must conduct heat from the die to the substrate, thermal conductivity is also a critical material property. The electrical properties of the attachment material are important when the attachment material functions as an ohmic contact providing an electrical path

Table 13. Electrical properties of attachment materials (Pecht 1991)

Material	Volume Resistivity (Ω-m)	Dielectric constant @1MHz	Dielectric strength (kV/mm)	Dissipation factor @1MHz
Silicone	10^{13}–10^{15}	2.9–4.0	15.8–27.6	0.001–0.002
Polyurethane	0.3×10^9	5.9–8.5	12.9–27.6	0.05–0.06
Acrylic	7×10^{11}	-	-	-
Epoxy novolak	10^{13}–10^{16}	3.4–3.6	-	0.016
Epoxy phenolic	6.1×10^{14}	3.4	15.8	0.32
Epoxy bisphenol A	10^{14}–10^{16}	3.2–3.8	-	0.013–0.024
Epoxypolyimide (50:50)	10^{12}	-	-	-

for electrical conduction from the device to the substrate circuit pad, as in circuit capacitors, chip transistors, and some resistors in multichip modules and hybrid packages. Electrically conductive attachment materials include silver-filled epoxies, silver-filled glasses and low-melting solders. Other required properties of attachment materials especially for manufacturing considerations include dispensibility, low curing temperature, wettability, low ionics and volatiles, and a uniform bonding line. Typical electrical, mechanical, and thermal properties of attachment materials are listed in Tables 13, 14, and 15, respectively.

Common die attach materials include organic adhesives such as epoxies or polyimides, gold-silicon eutectics, solders, and filled glasses. Metallurgical attachment is generally used for high-power circuits that require high thermal dissipation or for circuits that are susceptible to moisture. Solder or metal alloys generally provide good thermal conduction, but often cannot be used because they are also electrically conductive, render rework difficult, and can result in wirebond or device degradation due to the high temperature required to melt the solder. In some cases, epoxies or glasses, filled with metal or with thermally conductive oxides (alumina, beryllia), are used as a compromise.

Organic adhesives are widely used attachment materials due to their low cost and ease of rework. Two popular types of adhesives are electrically conductive and electrically insulative. Silver-filled or gold-filled conductive adhesives are commonly used to attach devices that require electrical conduction. Currently, silver-filled adhesives dominate the market because of their low cost. Integrated circuit dies that require no ohmic contacts are generally attached with nonconductive adhesives.

Generally, organic adhesives are not used for ceramic packages because the higher temperature needed to produce a frit seal after the die attach process may degrade the properties of the adhesive. Many commercial hermetic devices are assembled in ceramic cases supplied with a glass bottom,

Table 14. Mechanical properties of attachment materials (Pecht 1991)

Material	Tensile strength (MPa)	Shear strength (MPa)	Elongation(%)	Modulus of elasticity (GPa)	Specific gravity	Hardness[a]
Silicone	10.3	-	100-800	2.21	1.02-1.2	20-90A
Urethane	5.5-55	15.5	250-800	-	1.1-1.6	10A-80D
Acrylic	12.4-13.8	-	100-400	0.69-10.3	1.09	40-90A
Epoxy silicone	-	11.7	-	-	-	-
Epoxy novolak	55.0-82.7	26.2	2-5	2.76-3.45	1.2	-
Epoxy bisphenol A	43-85	-	4.40-11.0	2.7-3.3	1.15	106RM
Epoxy, electrically conductive	3.4-34	-	-	-	-	-
Modified polyimide	-	-	-	0.275	-	-
Polyimide	-	16.5	-	3.0	-	-
Epoxy polyimide	-	41	-	-	-	-
Epoxy polyurethane	-	-	10	-	-	-

[a] A = Shore A; D = Shore D; RM = Rockwell M

where electrical contact to the back side of the die is not needed. One formulation known as DIP 6 is popular in application on large dies. In this material, silver-filled glass is supplied as an 81% solid paste of silver-flake-loaded glass in an organic medium. The silver-glass attach layer can be kept practically void-free with ordinary care.

Gold-based eutectics like Au-Si, Au-Sn, and Au-Ge have rather high flow-stresses (onset of plastic flow) and therefore offer excellent fatigue and creep resistance. The disadvantage is primarily due to their lack of plastic flow, which leads to high stresses in the semiconductor due to the thermal expansion mismatch between the die and the substrate (Shukla and Mecinger 1985). The high cost also makes the gold-based eutectics less popular. Soft solders are usually tin-, lead-, or indium-based (e.g., 95Pb-5Sn and 65Sn-25Ag-10Sb, also called J-alloy) and are widely used. They are low in cost and possess lower melting points than hard solder. The possibility for oxide formation is reduced and lower die-attach temperatures are achievable, thereby reducing the chance of thermally degrading sensitive devices.

Die attach materials also include organic adhesives, such as epoxies or polyimides, filled with precious metals, hard solders (98Au/2Si, 80Au/20Sn, and 88Au/20Ge), soft solders (Pb-Sn and Pb-Ag-In alloys), and solders filled with glasses. Solventless conductive epoxies are the materials in widest use commercially, comprising about 80% of the die-attach market. They are normally filled with 70 to 80% silver, which imparts electrical conductivity and enhances thermal conductivity. There may also be coupling agents on the metallic particles to promote wetting and electrical conductivity. Formulations are available with or without an alumina filler, depending on whether electrical isolation of the die from the paddle is desired.

Table 15. Thermal properties of attachment materials (Pecht 1991, Minges 1989)

Material	Thermal conductivity (W/m-°C)	CTE (ppm/°C)	Heat capacity (J/kg-°C)	Maximum use temperature (°C)
Silicone	6.4–7.5	262–300	-	260
Polyurethane	1.9–4.6	90–450	-	65.6
Acrylic	-	-	-	93.3
Epoxy, silicone	13–26	60–80	-	260
Epoxy, phenolic	25.0–74.7	33	1674–2093	87.8
Epoxy, electrically conductive	0.17–1.5	-	-	-
Cyanoacrylate	-	-	-	82.2
Polyimide	0.2	40–50	-	-
Modified polyimide	-	73	-	-

ZEROTH-LEVEL PACKAGING MATERIALS

Table 16. Properties of Ablebond 8700E

Property	Numerical Value	Test Method
Thermal Conductivity @ 121°C	2.0772W/m°C	PT–40
Glass Transition Temperature (Tg)	160°C	MT–9
CTE below Tg	45ppm/°C	MT–9
above Tg	12ppm/°C	
Volume Resistivity	2.0×10^{-4} Ω-cm	PT–47
Die shear strength (Au to Au @25°C)	35163.48kPa	MT–4

These die-attach materials usually require a separate cure at 150 to 180°C for about an hour, although newer materials that cure in 5 to 10 minutes have also been developed.

Examples of a new epoxy adhesive and thin film are Ablebond 8700E and Ablefilm 5025E developed by Ablestik Electronic Materials and Adhesives. The properties of these materials are given in Tables 16 and 17.

Ablebond 8700E is an electrically conductive epoxy adhesive designed for the attachment of hybrid components. Suppliers' tests indicate that this adhesive provides good attachment to component termination surfaces, such as gold, without forming fissures or cracks. It retains its high shear strength even after thermal cycling, making it a suitable material for power applications in which heat is dissipated in relatively large quantities. This adhesive is also engineered to provide good dispensing and screen printing characteristics without tailing or peaking.

Ablefilm 5025E is a silver-filled unsupported epoxy adhesive film designed to provide uniform, thin bond lines. It provides excellent isotropic, electrical conductivity and good thermal conductivity. These properties make it suitable for microwave and heat sink applications.

Table 17. Properties of Ablefilm 5025E

Property	Unit	Test method
Lap Shear Strength		
Al to Al @25°C	17237 kPa	TLS – 1
Au to Au @25°C	20684.4 kPa	
Volume Resistivity	0.0002 Ω-cm	VR – 1
Water Extract Conductivity	15 μmhos/cm	EIA – 2
Glass Transition Temperature (Tg)	90°C	TG – 1
CTE Below Tg	65ppm/°C	TCE – 1
Above Tg	1.5ppm/°C	
Weight Loss @ 300°C	0.60 %	TGA – 1
Thermal Conductivity @121°C	3.462W/m°C	TC – 1

It provides radio frequency and EMI shielding when bonding microwave substances into packages. The high thermal conductivity is desirable for bonding "hot" devices onto heat sinks in applications where electrical insulation is not required.

2.3   Substrate Materials

Substrate materials can be broadly classified into single-layer and multi-layer materials depending on their material compositions. In hybrid packages, the substrate materials are classified into two categories, depending on the technology used for conductor metallization; the thick-film technology substrate materials and the thin-film technology substrate materials. For comparison, Tables 18 to 22 list the most common thin- and thick-film substrate materials and their properties

Substrate materials are required to meet various electrical, thermal, physical, and chemical requirements. Generally, a low dielectric constant is the most important property because parasitic capacitance effects are directly proportional to the dielectric constant. A low dissipation factor is required to reduce the electrical losses in the substrate material, particularly at high frequencies. High-volume resistivity is required to provide high electrical insulation to prevent electrical leakage current between the conductor tracks. High thermal conductivity is required to dissipate heat produced by the active devices on the die. An appropriate coefficient of thermal expansion is required to match the expansion coefficients of the die or other elements thus minimizing the thermal-mechanical stresses in the package during thermal and power cycling. High thermal stability is required to withstand the high temperature involved in manufacture, such as eutectic die attach and belt-furnace sealing. In addition, a high degree of surface smoothness is required, especially for hybrid packages and multi-chip modules, to achieve stable, precision, thin-film resistors and very fine conductor lines and spacings. The surface finish also affects the circuit loss and adhesion of films. Surface flatness is required to minimize problems during the film screening and photoprocessing used in thick-film technology. Non-flat surfaces do not press uniformly against screen printers or photomasks, resulting in non-uniform ink deposition or etched components. This can potentially lead to microcracking in the film components during substrate mounting into the package. Thin-film technology requires substrates with a smoother surface finish, as compared with the thick-film technology substrates.

Finally, the substrate should exhibit a high chemical resistance to withstand processing chemicals, such as the acids used in etching and plating, and low porosity and high purity to avoid moisture and contaminant entrapment, arcing, tracking, and metal migration.

Single-layer substrates: For single-layer substrates, a combination of specialized requirements quickly narrows the number of acceptable materials.

Table 18. Mechanical properties of substrates (Pecht 1991, Lide 1990, Kumar and Tummala 1991)

Materials	Tensile strength (MPa)	Compressive strength (MPa)	Flexural strength (MPa)	Elastic modulus (GPa)	Density (kg/m^3)	Hardness[a]	Impact strength (J)
Ceramics:							
BeO	230	-	250–490	345	3000	100K	-
SiC	17.24	490	440–460	412	3160	2800K	65
AlN	-	392–441	360–490	310–343	3260	1200K	-
Si	-	-	580	190	2330	850K–950K	-
Si$_3$N$_4$	96.5–965	-	275–932	314	2400–3440	-	-
SiO$_2$	96.5–386	-	30–100	69	2190–2320	-	-
Al$_2$O$_3$ 85%	124.11	1620	290	221	3970	9MH	8.5–8.8
Al$_2$O$_3$ 90%	137.9	2413	317	269	3970	9MH	8.8
Al$_2$O$_3$ 92%	127.56	1931	321	290	3970	-	8.8–9.2
Al$_2$O$_3$ 95%	127.4–193	2069–2413	310–338	296–317	3970	9MH	8.8–10.3
Al$_2$O$_3$ 96%	127.4	2344	317	310.3	3970	2000K	9.5
Al$_2$O$_3$ 99%	206.9	2586	345	345	3970	9MH	9.5
Al$_2$O$_3$ 99.5%	206.9	2620	345	379	3970	-	8.1
Al$_2$O$_3$ 99.8%	206.9	2758	350–414	386	3970	93.5RA	9.5
Al$_2$O$_3$	17.24	276	62	55.2	3970	6MH	4.5
Diamond	-	-	71.4	-	3500–3530	7500K	-
Quartz	48.3	1103	-	71.7	2200	5MH	-
Steatite	55.2–69.0	448–896	110–165	90–103	2500–2700	-	0.4–0.5
Forsterite	55.2–69.0	414–690	124–138	90–103	2700–2900	-	0.04–0.05
Titanate	27.6–69.0	276–827	69–152	69–103	3500–5500	-	0.4–0.7
Cordierite	55.2–69.0	138–310.3	10.3–48	13.8–34	1600–2100	-	0.3–0.34
Mullite	-	-	125–275	175	-	-	-
Metals:							
Molybdenum	655	-	-	324	10240	427K	-
Cu:W (10:90)	489.5	-	1062	331	17300	485K	-
Tungsten	310–1517	-	28.3	345	19300	68RB	-
Kovar	522–552	-	-	138	8360	-	-

a: K = Knoop; MH = Moh; RA = Rockwell A; B = Brinnell

Table 19. Thermal properties of substrates
(Pecht 1991, Lide 1990, Kumar and Tummala 1991)

Material	Thermal conductivity (W/m-°C)	CTE (ppm/°C)	Heat capacity (J/kg°C)	Maximum use temperature (°C)	Melting point (°C)
Ceramics					
BeO	150–300	6.3–7.5	1047–2093	-	2725
SiC	120–270	3.5–4.6	675	>1000	3100
AlN	82–320	4.3–4.7	745	>1000	2677
Si	125–148	2.33	712	-	1685
Si_3N_4	25–35	2.8–3.2	691	>1000	2173
SiO_2	1.5	0.6	-	>800	-
Al_2O_3	15–33	4.3–7.4	765	1600	2323
Quartz	43	1.0–5.5	816–1193	1140	1938
Diamond	2000–2300	1.0–1.2	509	-	N/A
Steatite	2.1–2.5	8.6–10.5	-	1000–1100	-
Fosterite	2.1–4.2	11	-	1000–1100	-
Titanate	3.3–4.2	7–10	-	-	-
Cordierite	1.3–4.0	2.5–3.0	770	1250	-
Mullite	5.0–6.7	4.0–4.2	-	-	-
Metals					
Molybdenum	138	3.0–5.5	251	-	2894
CuW (10/90)	209.3	6.0	209	-	-
Tungsten	174–177	4.5	132	-	3660
Kovar	15.5–17.0	5.87	439	-	1450

Materials commonly used as substrates include alumina, beryllia, silicon carbide, aluminum nitride, sapphire, glass, silicon, and quartz.

Alumina, or aluminum oxide, is by far the most commonly used ceramic for substrate materials (Seraphim et al. 1989). The composition varies between 90 to over 99% alumina, the balance being silicon dioxide, magnesium oxide, and calcium oxide. Alumina substrates used for thick-film technology are usually in the range of 94 to 96%. Higher purity alumina substrates (99.9% purity) with highly polished or glazed surfaces are commonly used as thin-film substrates. Alumina possesses the advantage of high electrical resistivity of the order of 10^{14} ohm-cm, which provides the substrate with excellent insulating characteristics. The dielectric constant is of the order of 9.5, which is quite low. The coefficient of thermal expansion is about 6.5 ppm/°C, which matches fairly closely that of GaAs. The grain size is approximately 3 to 5 μm and the surrounding glassy grain boundaries react with the thick-film binder glasses, giving significantly higher bond strength compared with other substrate materials.

Beryllia, or beryllium oxide (BeO), is often used as a substrate material in power hybrids. The thermal conductivity of BeO is approximately eight times that of alumina. However, beryllia is weaker than alumina and more expensive to produce in substrate form. Beryllia is also highly toxic in both powder and vapor form and requires special handling and safety equipment when machining or firing at high temperatures.

Aluminum nitride is a comparatively lower-cost alternative ceramic material that can be used instead of beryllia. Its thermal conductivity can be as high as that for beryllia, and in addition aluminum nitride provides closer matching of coefficients of thermal expansion with silicon, reducing residual stresses in the die attach.

Steatite and fosterite have been used for thick-film circuits. They have a lower dielectric constant than alumina (6.1 and 6.4, respectively, compared with 9.0 for 96% alumina) but have lower tensile. strength and thermal conductivity. Stealite and fosterite have often been used when the advantages of lower dielectric constant and usually lower cost have outweighed the disadvantages.

Table 20. Electrical properties of substrates
(Pecht 1991, Lide 1990, Kumar and Tummala 1991)

Material	Resistivity (Ω-m)	Dielectric constant @ 1MHz	Dielectric strength (kV/mm)	Dissipation factor @ 1MHz (10^{-4})
Ceramics				
BeO	10^{11} - 10^{12}	6.7-8.9	0.78	4-7
SiC	$>10^{11}$	20-42	0.04	500
AIN	$>10^{11}$	8.5-10	0.55	5-10
Si	640	11.9	-	-
Si_3N_4	$>10^{10}$	6-10	196.8	-
SiO_2	$>10^{14}$	3.5-4.0	196.8	-
Al_2O_3	10^9-10^{12}	4.5-10	0.33	2-100
Diamond	$>10^4$	5.7	-	-
Quartz	-	4.6	-	<4
Steatite	10^{11}-10^{13}	5.5-7.5	7.9-15.7	2-40
Forsterite	10^{10}-10^{12}	6.2	7.9-11.8	3-4
Titanate	10^6-10^{13}	15-12000	2.0-11.8	2-500
Cordierite	10^6-10^{14}	4.5-5.5	1.6-9.8	40-120
Mullite	$>10^{12}$	6.6-6.8	-	-
Metals				
Molybdenum	$(4.9\text{-}5.2)\times10^{-8}$			
CuW (10/90)	6.6×10^{-8}			
Tungsten	$(4.8\text{-}5.5)\times10^{-8}$			
Kovar	50×10^{-8}			

Glass substrates are used in low-power, low-frequency, and high-precision circuitry. Quartz and sapphire substrates have extremely low loss compared with other substrates and have found extensive use in microwave thin-film circuits. However, glass is a very poor thermal conductor compared with alumina. Sapphire has a much higher thermal conductivity than quartz and provides a good thin-film substrate for higher-power microwave circuits. Both sapphire and quartz are very costly and are not machined as easily as alumina.

Multilayer cofired materials for hybrid substrates: The number of dielectric layers in a multilayer substrate can vary between two and forty, depending on the application. Each layer is typically electrically connected to its adjacent layers through vias. Depending on the material constituents and sintering temperature of the substrates, the multilayer ceramic substrates can be classified as high-temperature cofired ceramic (HTCC) or low-temperature cofired ceramic (LTCC) systems (Table 23).

High-temperature cofired ceramics have a high ratio of ceramic to glass as listed in Table 23. The green substrate layer can only be sintered at firing temperatures of approximately 1500°C. In such material systems the dielectric consists of glass fillers in a ceramic matrix. Consequently, thick-film pastes that are cofired with the substrate layer also have to withstand these high temperatures. Usually, tungsten and molybdenum-manganese are used as the metallization materials.

Table 21. Material properties of key thick film substrates (Licari and Enlow 1988)

Properties	96% alumina	99.5% alumina	99.5% beryllia
Dielectric constant			
@ 1 MHz	9.3	9.9	6.9
@ 1 GHz	9.2	9.8	6.8
Dielectric strength (kV/mm)	8.3	8.7	9.1
Dissipation factor			
@ 1 MHz	0.0003	0.0001	0.0002
@ 1 GHz	0.0009	0.0004	0.0003
Thermal conductivity (W/m-K)			
@ 25°C	35.1	6.7	250
@ 300°C	17.1	18.7	121
CTE @25°C-300°C (ppm/°C)	6.4	6.6	7.5
Bulk resistivity (10^{13} Ω-cm)			
@ 25°C	10	10	10
@ 100°C	2	7.3	10
Tensile strength (MPa)	172.4	193.1	158.6
Surface finish (μm)	0.10	0.10	0.08

Low-temperature cofired ceramics have a much lower fraction of ceramic compared to HTCC. LTCC can be sintered at much lower firing temperatures (850–900°C). Low electrical resistivity on conductors is obtained by controlling the silver and palladium particle shape. A reduction in the processing temperature to around 900°C facilitates the use of low-resistivity metals such as silver-palladium, silver, gold or copper. The LTCC substrate is advantageous for high wiring density due to decreasing line dimensions such as width and thickness, and for high-frequency pulses because of decreasing waveform rounding on the rise pulse.

Gold conductors can be used for interconnections in LTCCs if cost is not a limitation. However silver-palladium achieves lower resistivity than gold at lower cost. The dielectric constant materials for LTCC substrates are 55 wt % alumina to 45 wt% lead borosilicate glass, 45 wt% cordierite - 55 wt% borosilicate glass, and 35 wt% silica - 65 wt% borosilicate glass, which have controlled particle size and dielectric constants of 7.8, 5.0 and 3.9, respectively. Silver-palladium wiring with low resistivity reduces voltage drops and localized heating, which can affect device switching behavior.

Table 22. Material properties of key thin film substrates (Licari and Enlow 1988)

Properties	Alumina	Beryllia	Glass	Quartz	Sapphire
Dielectric constant	10.1 (1.0 GHz)	6.9 (1 MHz)	5.84 (1.0 GHz)	3.83 (1 MHz)	9.39 (1 GHz)
	10.7 (9.9 GHz)	6.8 (1 GHz)	5.74 (8.6 GHz)	3.82 (6 GHz)	9.39 (10GHz)
Dielectric strength (kV/mm)	30.3	9.1	-	16.1	190
Thermal conductivity @ 25°C (W/m-°C)	36.7	250	1.7	1.4	41.7
CTE @ 25°C (ppm/°C)	6.7	7.5	4.6	0.49	-
Bulk resistivity @ 25°C (Ω- cm)	3.16×10^{10}	10^{14}	10^{14}	31.6×10^9	10^{14}
Thickness available (mm)	0.254-1.016	0.635	0.813 & 1.626	-	-
Tensile strength (MPa)	-	158.6	-	48.3	399.9
Surface finish (μ m)	1.0	15-20	1.0	1.0	1.0

Table 23. Typical constituents of high- and low-temperature cofired ceramic systems (Neugebauer et al. 1991)

Construction	High-temperature cofired	Low-temperature cofired
Dielectric		
Matrix	Ceramic (96 wt% alumina)	Glass/crystallizable glass
Filler	Glass (4 wt% glass)	30 to 50 wt% ceramic
Organic binder	Poly vinyl butryl	Acrylates/methacrylates
Solvent	Toluene	Xylene
Plasticizer	Polyethylene glycol	Dicotyl phthalate
Metallization	Tungsten, moly-manganese	Ag, Au, Cu, Ag-Pd

Low epsilon ceramic (LEC) is composed of alumina and quartz, with borosilicate as the base glass. LEC's greatest advantage over high temperature ceramics is its low dielectric constant, which allows for high-speed transmission lines.

High-density substrate materials for multichip modules (MCMs): In the application of multichip modules, high-density substrates are used. High-density substrates can be achieved in two ways: extending standard ceramic lamination technologies which results in many layers of relatively large geometries, or extending standard thin-film technology, which results in few layers of very small geometry. Table 24 lists the five basic categories of high density substrates, including MCM-L, MCM-C, MCM-D, MCM-D/C, and MCM-Si as well as some features associated with the substrate materials.

MCM-L encompasses substrates fabricated using reinforced and unreinforced organic dielectric pc-board materials and processes. The typical materials are epoxies and polyimides with thick copper lines formed by either additive or subtractive methods. The metallized dielectric layers are laminated together. Interlevel connections are typically formed by drilling holes through the layers and metallizing them using electroless (a chemical reduction process, see Section 4.6) and electroplated copper. The pc-board infrastructure is well established, making the MCM-L a low-cost alternative for many MCM applications.

The main advantages of MCM-L are excellent electrical properties, low cost, area-array I/O and the ability to produce solid vias. Solidly filled vias offer low resistance and can be used to construct solid posts of copper running through the substrate to conduct heat. The electrical properties are good because copper is an excellent conductor and the organics, such as polyimide, have very low dielectric constants. Cost is low because large sheets of material, each containing multiple substrates, can be processed at one time.

MCM-C ceramic substrate (cofired ceramic) consists of ceramic layers for power and ground planes and thin layers for high-density routing of signals. MCM-C substrates offer the advantage of area-array I/O off the bottom of the substrate, an important feature for modules with many I/Os.

Table 24. High density substrates and selected attributes (Blood and Casey 1991)

Substrate[*]	MCM-L	MCM-C	MCM-D	MCM-D/C	MCM-Si
Description	laminated high density PCB	cofired ceramic	deposited organic thin film on Si, ceramic or metal	deposited thin film on cofired ceramic	SiO_2 dielectric with Si substrate
Maturity	good	very good	limited	limited	limited
Cost	medium	medium	high	high	medium
Number of metal layers	>15	>50	5	>50	5
Minimum metal pitch(μ m)	100–150	250–450	25–75	50–75	25
Substrate I/O	array	array	peripheral	array	peripheral
Heat transfer	fair	poor	good	poor	good
Dielectric constant	3.0–3.5	9.7	<3.0	<3.0	3.5

[*] MCM - Multi Chip Module; L - Laminate; C - Ceramic; D - Deposit; SiO_2 - Silicon dioxide

These substrates can be easily designed as a hermetic package using a built-in seal ring around the periphery. MCM-Cs using high-temperature cofired ceramic are characterized by high line resistance due to the low conductivity of tungsten. MCM-C works well for high-I/O, medium-performance modules such as 50MHz to 100MHz processor clocks. MCM-Cs with glass-ceramic dielectric materials fabricated with LTCC technology have low dielectric constants, low coefficients of thermal expansion (CTEs), and compatibility with low-resistivity conductor materials at low firing temperatures. Cost effectiveness remains questionable because the conductor screen printing process limits the achievable interconnect density.

MCM-D consists of a substrate deposited with thin film on silicon, ceramic, or metal. MCM-D substrates are used in applications that require high electrical performance and high interconnection densities with a minimum number of substrate layers. The thin-film processing is accomplished on a rigid base material, usually silicon, alumina, or metal. Commonly used thin-film materials in MCM-D include lower-conductivity aluminum and organic dielectric materials because the processing is easy and reliable. Copper is used sometimes for its better conductivity. However, there is a reliability problem when uncured polyimide comes into contact with copper. The problem can be eliminated by adding barrier metals such as chromium or nickel.

MCM-Si substrates use a silicon wafer, with a deposited thin-film of silicon dioxide as the dielectric, and aluminum or copper as the conductive materials. Small geometries, improved reliability (over organic dielectrics), the ability to incorporate decoupling capacitance (up to 42nF/cm sq) in the substrate, and high thermal conductivity of the substrate are the major advantages. Additionally, the coefficient of thermal expansion match of a silicon substrate to silicon chips is a great advantage.

MCM-D/C, with deposited thin film on cofired ceramic technology, is currently being produced only in Japan. It offers the best of ceramic and thin-film technologies and is an ideal choice for all types of modules. High cost is perhaps its only disadvantage.

3 FIRST-LEVEL PACKAGING

First-level (or component level) packaging is designed to interconnect the device to the board while providing protection for the device against mechanical stress and chemical attack. In addition, in advanced packaging structures, first-level packaging can perform some of the signal processing and distribution functions as well. In order to accomplish these tasks, the package consists of a number of key elements described in further detail here, including wirebonds, TAB attach, cases, leads, and lid seals.

3.1 Wire Interconnects

Wirebonding is the most common technique for interconnecting a silicon chip to a chip carrier or to the leadframe in a single-chip package. Gold and aluminum are the usual wire materials, although copper and silver have also been used. Gold is mostly used in small signal devices because of its ability to be drawn into small-diameter wires (1 mil) and to create non-directional thermocompression ball bonds. Ultrasonically bonded aluminum wire, which is less expensive, is mostly used in power devices, where thicker wires are needed and where a monometallic bond is preferred for its immunity to the formation of intermetallics. Critical material properties of wire and bond-pad material include shear, yield, and ultimate strength, elastic modulus, Poisson's ratio, hardness, CTE, elongation, fracture toughness, and fatigue endurance in tension and shear. Of particular importance to the strength of the wirebond are wire dimensions, tensile strength, and elongation. The properties of various wire and bond-pad materials are listed in Table 25.

Gold bond-wire systems Gold wire is used extensively for thermocompression bonding, although either thermocompression or thermosonic bonding can be used with gold. In producing gold bonding wires, control of the surface finish and surface cleanliness are of the greatest importance in ensuring the formation of a strong bond and preventing the clogging of bonding capillaries. Pure gold can usually be drawn to an adequate breaking strength (the ultimate tensile strength of the wire) and proper elongation (the ratio of the increase in wire length at rupture to the initial wire length given as a percent). Ultra-pure gold is very soft, but even after the addition of small amounts of impurities such as 5 to 10 ppm by weight of beryllium or 30 to 100 ppm by weight of copper, the gold is still ductile. Beryllium-doped wire is stronger than copper-doped wire by about 10–20% under most conditions. The increased strength of the Be-doped wire is advantageous for automated thermocompression bonding, where high-speed capillary movements generate higher stresses than in manual bonders.

The most common wire-bonding system in commercial electronic packages is gold wire bonded to aluminum bond pads. This takes advantage of the ease of drawing gold into narrow-diameter wires, gold's resistance to corrosion and high conductivity, and the ability to make non-directional

Table 25. Properties of wire and bond pad materials @ 27°C

Property	Aluminum (Al)	Copper (Cu)	Gold(Au)
Specific heat (W-sec/g-°C)	0.9	0.385	0.129
Thermal conductivity (W/cm-°C)	2.37	4.03	3.19
Specific gravity	2.7	9.0	19.3
Melting point (°C)	660	1083	1064
Electrical resistivity (Ω-cm)	2.65×10^{-6}	1.7×10^{-6}	2.2×10^{-6}
Temperature coefficient of electrical resistivity (Ω-cm/°C)	4.3×10^{-9}	6.8×10^{-9}	4×10^{-9}
Elastic modulus (GPa)	34.5	1324	77.2
Yield strength (MPa)	10.34	68.95	172.38
Ultimate tensile strength (MPa)	44.82	220.64	206.85
Coefficient of thermal expansion (ppm/°C)	46.4	16.12	14.2
Poisson's ratio	0.346	0.339	0.291
Hardness (Brinnell)	17	37	18.5
Elongation (%)	50	51	4

interconnects through ball bonding. On the bond-pad side, it takes advantage of the ubiquity of aluminum metallization and therefore the lack of any additional required device processing steps to produce the bond pads. The disadvantage is that this system is highly susceptible to Kirkendall voiding as a result of dissimilar interdiffusion rates, and to the formation of a number of brittle intermetallic compounds at temperatures just above the range of normal microcircuit operation (125°C). The combination of voids and brittle interfacial layers significantly increases the electrical resistance of the bond and decreases its mechanical fatigue resistance over time at temperature. As a result, this wirebonding system is not used for any applications above 125°C. Another drawback is the limitation of the amount of current that can be carried by the slim gold wires. This inhibits the use of gold wires in power electronics, where thicker, lower-cost aluminum wires are used.

Gold wires can be bonded to copper leadframes. However, this can also lead to the formation of intermetallic phases (Cu_3Au, $AuCu$, Au_3Cu) in the neighborhood of 200°C to 325°C, which, when combined with the creation of stress-concentrating voids by interdiffusion, can decrease the bond strength (Hall et al. 1975). Temperatures above 300°C accelerate the formation rate of intermetallic compounds. However, Pitt and Needes (1982) studied gold thermosonic bonds to thick-film copper and found little strength degradation at 150°C for up to 3000 hours and no failures at 250°C for over 3000 hours. Cleanliness of the bonding surface is extremely important to ensure good bondability in Cu-Au systems (Lang and Pinnamenni 1988; Fister et al. 1982).

Gold bonded to silver metallization is very reliable for a long time at temperatures in the range of 155°C to 450°C (James 1977). This bond system does not form intermetallic compounds and does not exhibit interface corrosion, although contaminants like sulfur reduce the bondability of silver films to gold. High-temperature (approximately 250°C) thermosonic bonding may be needed to increase the bondability of silver (Harman 1989).

Gold, wire-bonded to a gold bond pad, is even more reliable because the bond is not subject to degradation due to interface corrosion, or intermetallic formation, since it is monometallic. Instead of degrading, even a poorly welded gold–gold bond will increase in strength with time and temperature (Jellison 1975). For this reason, gold–gold bonds are preferred in high-temperature electronic systems (McCluskey 1996). Failures in the gold–gold system at high temperature are not related to bond pad degradation but to wire fatigue resulting from annealing of the gold wire (Grzybowski 1996).

Aluminum bond-wire systems: Pure aluminum is typically too soft to be drawn into a fine wire. For this reason aluminum is often alloyed with 1% Si or 1% Mg to provide a solid-solution strengthening mechanism. However, the equilibrium solid-state solubility of silicon in aluminum at 20°C is of the order of 0.02% by weight, and only at temperatures above about 500°C does silicon attain 1% solubility in equilibrium solid solution. Thus 1% silicon exceeds the solubility of silicon in aluminum at room temperature by a factor of 50. As a result, at ordinary room temperatures, there is always a tendency for a second silicon-rich phase to precipitate. The silicon-rich precipitates formed by this precipitation hardening mechanism can lead to potential fatigue crack nucleation sites.

Aluminum alloyed with 1% magnesium can be drawn into a fine wire that exhibits a breaking strength similar to that of Al-1%Si. The Al-1%Mg alloy wire bonds satisfactorily and is superior to Al-1%Si in resistance to fatigue and to loss of ultimate strength after exposure to elevated temperatures. These advantages occur because the equilibrium solid solubility of magnesium in aluminum is about 2% by weight and thus, at 0.5 to 1% magnesium concentration, there is no tendency towards second-phase segregation, as is the case with aluminum-1%silicon.

Olsen and James (1984) observed that for aluminum wire bonded to oxygen-free high-conductivity (OFHC) copper, such as is present in a leadframe, thermal aging in a 150°C ambient did not affect the bond strength. However, thermal aging in a vacuum resulted in a weaker bond.

Aluminum wire-bonding to a silver-plated leadframe is often used in thick-film hybrids. The Ag–Al phase diagram is very complex with many intermetallic phases.   Kirkendall voids occur in this metal system, but typically at temperatures greater than 200°C, thus higher than the operating range of most microcircuits. Ag-Al bonds are seldom used because of their tendency to degrade due to interdiffusion and to oxidize in the presence of humidity. Al–Cl corrosion is responsible for significant bond degradation as indicated by the formation of aluminum hydroxide, $Al(OH)_3$, in the zeta phase of the Ag-Al intermetallic (Hermansky 1972; Jellison 1975; Kamigo and Igarashi 1985; James 1977).

Harman (1989) found that aluminum wire-bonding to nickel coatings, which are used as a substitute for gold, is reliable under various environments, more so than Al–Ag or Al–Au bonds. Large diameter (>75 mm or 3-mil) aluminum wires bonded to nickel platings have been used in power devices (Harman 1989). The Al–Ni system is well suited to high-temperature applications and is less prone to Kirkendall voiding and galvanic corrosion than other bimetallic systems.

Aluminum wire-bonding to aluminum metallization is extremely reliable because it is not prone to intermetallic formation or corrosion, since it is monometallic. As with Au–Au, this immunity to corrosion and intermetallic formation makes Al–Al an excellent choice for high-temperature electronics (McCluskey 1996). Also similarly to Au–Au, failure occurs in the wire not at the wirebond (Grzybowski 1996). Aluminum wire on aluminum metallization welds best ultrasonically, even though a thermocompression bond can be produced by high deformation.

Copper bond-wire systems: Copper wires are used primarily because of their low cost and their resistance to sweep (the tendency of the wire to move in the plane perpendicular to its length) during molding when used in plastic-encapsulated components (Hirota et al. 1985; Atsumi et al. 1986; Levine and Shaffer 1986; Onuki et al. 1987 and Riches and Stockham 1987). Copper is harder than gold and thus greater attention is needed during the bonding operation to prevent cratering of the silicon die, as the harder copper wire tends to push the softer metallization aside during the bonding operation. Bonding systems using copper wires therefore require harder metallization (Riches and Stockham 1987; and Hirota et al. 1985). Bonding systems using copper wires also require an inert atmosphere for bonding because of the tendency of copper to oxidize readily. Intermetallic growth in Cu-Al bonds is slower than Au-Al bonds. The intermetallic growth in Cu-Al bonds does not result in Kirkendall voiding but produces a lower shear strength at 150–200°C due to growth of a brittle $CuAl_2$ phase (Atsumi et al. 1986 and

Table 26. Properties of bump materials used in TAB

Bump Material	Tensile Strength (MPa)	Yield Strength (MPa)	Shear Strength (MPa)	Elongation (%)	Elastic Modulus (GPa)	Thermal Conductivity (W/m-K)	CTE (10^{-6}/°C)	Heat Capacity (J/kg K)
Au (0.1% Co, cyanide) (1%Co,cyanide)						318	14.2	129
Au (cyanide)	152–213	88–128		3.5–6.3		318	14.2	129
Cu (cyanide)	441–618			4–18	108–117		16.7	
Cu (pyrophosphate)	280			35	117		16.7	
Cu (sulfate)	137–265			8–41	108		17.1–18.9	
Ni, electroless	441–771			3–6	117–196	4.3–5.7	13.5–14.5	
95Pb/5Sn	23.2	13.3		26.0	23.5	35.5	28.7	
37Pb/63Sn	35.4–42.2	16.1	28.5	1.38	14.9	50.6	24.7	
51In/32.5Bi/16.5Sn								
50Pb/50In	32.2		18.5	55			26.3	
95Pb/5In	29.8		22.2	52				

Onuki et al. 1987). Cu-Al bonds are stronger than Au-Al bonds in the presence of brominated flame retardants. Thomas et al. (1977) report on a study where Cu-Al bonds were strong after 1245 hours at 200°C, but Au-Al failed after 700 hours. Aluminum metallization containing copper-aluminum intermetallics is susceptible to corrosion when exposed to chlorine contamination and water.

Bonding surfaces: Bonding surfaces used on the die include aluminum, gold, silver, nickel and copper. Aluminum is the most commonly used material, although gold is used to avoid direct contact between gold and aluminum to prevent the formation of gold intermetallics. Because gold does not adhere well to silicon dioxide, other metals are used to form a multilayer metallization to avoid direct contact between gold and silicon.

Silver is used as a bond plating material on leadframes and as a metallization in commercial thick-film hybrids, usually in alloy form with platinum or palladium (Baker et al. 1981). At temperatures in the neighborhood of 400°C, silver undergoes selective oxidation and is therefore not used in high-temperature applications.

Nickel has been widely used in power devices as a substitute for gold in various environments with no significant reliability problems. Nickel can be deposited from electroless boride or sulfamate solutions, with the low-stress films electroplated from sulfamate baths resulting in reliable bonds. Bonding to the die and the terminal is affected by many film-related factors, including surface finish, film hardness, film thickness, film preparation, and surface contamination. Nickel surfaces are prone to oxidation and should be bonded soon after being nickel-plated, chemically cleaned before bonding, and protected by an inert atmosphere (Harman 1989). Phosphide electroless nickel solutions that deposit films containing 6-8% of phosphorus can result in reliability and bondability problems.

3.2   Tape Interconnects

The basic elements which constitute tape automated bonding (TAB) include the tape, the metal plating, the interface metallurgy (IFM), the bump, and the adhesion layer. Each of these elements requires materials with their own unique properties. Table 26 lists properties of common materials used in TAB (Lau et al. 1989 and Pecht 1991).

Tape: The tape materials should have dimensional stability, the capability to withstand the short-term elevated temperature required for solder reflow and thermocompression bonding, good surface flatness, non-flammability, low shrinkage, low moisture absorption, a high tensile modulus, a low CTE a high dielectric constant, and good ability to withstand exposure to plating, etching, and soldering.

The high tensile modulus is desirable because materials with a higher modulus (greater stiffness) are less prone to "curl" or become non-planar. Two-layer adhesiveless tape is often prone to curling, because it is not stiff enough. Adhesives and thicker dielectric layers are often used to solve the problem.

During bonding, the tape can shrink and create residual stress at bonding points. A lower coefficient of thermal expansion means less change in the tape's dimensions during the bonding process and a lower likelihood of distortions that can create residual stress. Low water absorption and a low coefficient of humidity expansion is preferable to avoid stresses related to swelling.

Choices for the tape material include polyimide, polyester, polyethersulfone (PES), or polyparabanic acid (PPA) (Lyman 1975). Among them, polyimide is most widely used because it can survive up to 365°C due to its high melting point. Polyester is limited to only 160°C but is much cheaper. PES and PPA are new materials and cost about half as much as polyimide. They can survive maximum short-term temperatures of 220°C and 275°C, respectively. PES has shown better dimensional stability than polyimide, absorbs less moisture, and does not tear as easily, but it is flammable and is attacked by some common solvents. The properties of tape materials are listed in Table 27.

Metal Plating: Metal plating should have excellent electrical and thermal conductivity, ductility (high toughness), adherence to the plastic carrier, and the capability to be easily etched. High electrical and thermal conductivity is desirable for electrical performance and heat dissipation. The high toughness associated with a fine-grain, equiaxial structure allows the material to withstand more thermal cycling stress during bonding and operation. It also makes processing easier.

Adhesion of the metal plating to the plastic tape is obtained by using adhesive layers for three-layer tape structures. For one-layer and two-layer tape structures, no adhesive is needed. Once adhered to the tape, the metal pattern is photolithographically defined and then etched.

Copper is used universally as the tape metal because of the ease of the lamination processes between copper and various plastic tapes, the ability of copper to be readily etched, and copper's excellent electrical and thermal conductivity, in both electro-deposited and rolled-annealed form (Section 4.6 contains additional information pertaining to the use of copper).

The tape metal is normally gold or tin-plated to ensure good bondability during inner lead bonding (ILB) and outer lead bonding (OLB) operations, as well as to provide long shelf life and corrosion resistance. Gold-plated copper leads are used for thermal compression and solder reflow bonding while tin-plated copper leads are used for eutectic soldering. Tin-gold eutectic soldering has been widely reported to have tin whisker problems and poor reliability (Dosworth 1991).

Interface Metallurgy: The chip pad is normally coated with an interface metallurgy (IFM) to provide low contact resistance, high resistance against corrosion, adhesion to the pad, and storage stability. The IFM material structure generally includes an adhesion layer, a diffusion barrier layer and a bonding layer. The adhesion layer is chosen to make good electrical contact to the pad as well as to adhere to the passivation layer surrounding the pad. The barrier layer prevents unwanted interdiffusion between the several metals of the bonding structure. The bonding layer is normally a noble metal to provide an inert surface for bonding or plating. For a bumped tape, this interface metallization is a bonding surface rather than a plating base. For a bumped chip, the bump is electroplated on the IFM and the chip interconnection pad is normally coated with several layers of metal to provide low contact resistance and adhesion to the pad. The top layer of the interface metallurgy is usually a noble metal to provide an inert surface for bonding and plating.

Common materials for IFM are titanium or chromium as an adhesion layer, copper, nickel, tungsten, palladium or platinum as a barrier layer; and gold as a bonding layer. Soft gold is desirable to accommodate thermal compression bonding. Some companies use a duplex copper/gold structure to minimize the amount of gold used. The solder bump in solder reflow TAB is usually made of 95% lead and 5% tin by weight. Such a high-lead solder has a 315°C melting point, higher than most other temperatures expected in the packaging process (Speerschneider and Lee 1989).

Bumps: Bump materials should be soft and ductile. Hard bump materials can lead to "silicon-cratering" during thermocompression bonding. To avoid cratering, the bumped wafer may be annealed prior to thermocompression bonding, or it can be bonded using eutectic soldering rather than thermocompression bonding. Solder-bump reflow TAB has a number of advantages over gold or copper bump TAB (Speerschneider and Lee 1989). First, solder-reflow TAB requires a very low bonding force, as the force is only required to hold the tape in place and to overcome the surface tension of the molten solder bump. Secondly, solder-reflow TAB is very forgiving with regards to planarity. The solder bump typically is about 100 mm in height at the start of bonding and 50 mm high after bonding. This allows a variation of about 50 mm in planarity for producing uniform bonds across the chip surface. For gold or copper bump TAB, experience has shown that planarity variation is limited to around 10 mm. The choice for the solder material is based on melting temperature and strength.

Adhesion: The adhesive or the adhesion layer that attaches the metal foil to the plastic carrier should be mechanically stable during thermal excursions to ensure good connection between metal and film. Adhesives have been used in three-layer tape, including polyimides, polyesters, epoxies, acrylics and phenolic-butyrals. They are rated for only 20 to 30 seconds at 200°C, and therefore limit the temperature at which the tape can be processed. Table 28 lists material properties of adhesives.

3.3   Case Materials

Package cases should be made of materials that protect the die circuitry and the interconnections to the leads from mechanical stresses and chemical attack. Also, the package cases should provide electrical insulation of the component. If the die(s) are mounted directly on the case header (or base), a low dielectric constant for the case material is desirable, as it results in low capacitive loading, thus minimizing the propagation delay. The case material should also have a high coefficient of thermal conductivity to facilitate the process of dissipation of heat generated during component operation, while the coefficient of thermal expansion of the package should match closely with the die, substrate, and sealing materials to avoid excessive residual stresses and fatigue damage under thermal cycling loads. Other requirements include high strength and toughness for processability, and stability at the processing temperatures used during fabrication.

Cases can be classified as hermetic or non-hermetic based on their permeability to moisture. Hermetic case materials, typically metals and ceramics, do not allow permeation of moisture through them. High-reliability devices used by the military have traditionally been packaged in hermetic cases to prevent chemical degradation of the die and the interconnections due to ionically contaminated moisture. On the other hand, commercial electronic devices are typically encapsulated in a silica-filled epoxy resin because of lower cost, reduced size and weight, and often superior electrical performance. As such a case is permeable to moisture, it is called a non-hermetic case. Recent studies have shown that high reliability can be achieved with a non-hermetic case if ionic contamination is limited.

The common hermetic case materials and their properties are listed in Table 29. One of the most widely used materials for the fabrication of metal cases, especially lids, is Kovar, a proprietary metal alloy composed of 54% Fe, 29% Ni and 17% Co. Kovar has a coefficient of thermal expansion of 5.1-5.9 ppm/°C; this closely matches those for the many commonly used sealing glasses, which vary from 5.25 to 6.96 ppm/°C (refer to Table 30). However, the thermal conductivity of Kovar is only 15.5-17 W/m-°C, which is low compared with that for copper alloys. As a result, the package case is not very effective in dissipating the heat generated during component operation and is not a suitable case material for high-power applications. In such applications, copper alloys are more suitable because of their high thermal conductivity. For example, the conductivity of a commonly used copper–tungsten alloy is 209.3 W/m-°C. However, it is very difficult to seal the lid in a copper case. Large thermal stresses are generated in sealing glasses because of the CTE mismatch, while electrical resistance welding (the preferred process for hermetic sealing) is difficult because the high thermal conductivity of copper causes the heat generated during the welding process to dissipate very quickly. A low-cost alternative that offers easy weldability is cold rolled steel (CRS), although it is more susceptible to corrosion.

Table 27. Properties of plastic films. (Lau et al. 1989)

Plastic film	Melting point (°C)	Moisture absorption (%)	Tensile strength (MPa)	Ultimate elongation (%)	CTE (10^-6/K)	Dissipation factor @ 1 kHz	Dielectric constant @ 1kHz	Dielectric strength (MV/m)	Cost factor
Polyimide (Kapton)	NONE	3.0	230	72	18–50	0.0018–0.0026	3.4–3.5	150–300	17.0
Polyester (Mylar)	180	<0.8	170	120	31	0.005	3.25	300	1.0
TFE (Teflon)	328	<0.01	28	350	12.2	0.0002	2.0	17	9.2
FEP (Teflon)	280	<0.01	21	300	9.7	0.0002	2.0	255	16.0
Polyamide (Nomex)	482	3.0	75	10	38–154	0.007	2.0	18	1.9
Polyvinyl chloride	163	<0.5	35	130	63	0.009	3.0	40	0.36
Polyvinyl fluoride (Tedlar)	299	<0.05	70–130	110–300	50	0.02	8.5–10.5	140	3.5
Polyethylene	121	<0.01	21	>300	198	0.0003	2.2	20	0.17
Polypropylene	204	<0.005	170	250	104–184	0.0003	2.1	160	0.48
Polycarbonate	132	0.35	62	110	68	1.000	3.2	16	1.32
Polysulfone	190	0.22	68	95	56	0.001	3.1	295	3.33
Polyparabanic acid (PPA)	299	1.8	97	10	51	0.004	3.8	235	8.0
Polyether sulfone (PES)	203		84			0.001	3.5		4.5

Table 28. Material properties of adhesives (Lau et al. 1989)

Adhesive	Type	Nip roll temperature ($^{\circ}$ C)	Maximum solder temperature ($^{\circ}$ C)	Minimum peel strength (N/mm)
Rogers 8145	Epoxy	204	316	0.53
DuPont WA	Acrylic	188	316	1.8
CMC 1477	Epoxy	190	288	1.58
CMCX-1496	Polyester	190	232	1.05

Ceramics are also extensively used for the construction of hermetic packages. Aluminum oxide or alumina is by far the most common ceramic used in electronic packaging (Seraphim et al. 1989). The composition varies between 90 and 99% alumina, the balance being silicon dioxide, magnesium oxide, and calcium oxide. Alumina possesses the advantage of high electrical resistivity of the order of 10^{12} ohm-cm, which gives the case excellent insulating characteristics. The coefficient of thermal expansion ranges between 4.3 and 7.4 ppm/$^{\circ}$C which closely matches that of silicon and many commonly used sealing glasses (refer to Table 30). However, the thermal conductivity of alumina is about 15 W/m-$^{\circ}$C, which compares unfavorably with that for copper alloys (for example, 209.3 W/m-$^{\circ}$C for copper-tungsten alloy). Hence, alumina cannot be used as the header material in high-power applications. In such applications, beryllium oxide, aluminum nitride, or silicon carbide is used.

Although BeO has a very high thermal conductivity (150–300 W/m-$^{\circ}$C), it is highly toxic in powder form and special precautions are required for the safety of workers during its processing. Beryllia is also about ten times as costly as alumina (Seraphim et al. 1989) and has a poorer CTE match to silicon. An alternative ceramic material that can be used instead of BeO is AlN because its thermal conductivity can be as high as 170–260 W/m-$^{\circ}$C while its coefficient of thermal expansion is 4.3–4.7 ppm/$^{\circ}$C (that of Si), reducing the residual stresses in the die attach.

Another category of materials for cases is the metal matrix composites currently under development. These materials are formed by introducing ceramic particles into a matrix of metal. A typical example is discontinuously reinforced aluminum (DRA), which is a composite of silicon carbide particles in an aluminum matrix. This material offers the advantage of light weight, a coefficient of thermal expansion close to that of common die materials, and thermal conductivities comparable to those of metals. The coefficient of thermal expansion for the silicon carbide-aluminum composite can be tailored (according to that of the die) from 6 ppm/$^{\circ}$C to 15 ppm/$^{\circ}$C by changing the percentage of silicon carbide in the composite (White 1990). The thermal conductivity of the composite is about 160 W/m-$^{\circ}$C as compared to about 15.5–17.0 W/m-$^{\circ}$C for Kovar and 15.33 W/m-$^{\circ}$C for alumina. Metal matrix composites combine the advantages of the low coefficient of thermal expansion of ceramics and the high thermal conductivity of metals and should play a larger role in future packaging.

3.4   Lid Seals

Lid seal materials are required to seal the lids of hermetic packages. Seal materials should have low permeability to moisture, chemical inertness, high shear strength, favorable fatigue properties, low sealing temperature, good wetting properties with respect to the case and lid surfaces, and a coefficient of thermal expansion that matches the lid and case materials. Seal materials are chosen to be virtually impermeable to moisture because moisture ingress into the package can contribute to corrosion of the device and interconnects. The seal material should also, in itself, be resistant to corrosion so that it retains its design strength and impermeability to moisture. The shear strength and fatigue properties (along with elastic constants) of the seal material are required to evaluate its ability to withstand service loads. Information about the sealing temperature of the material used is required because the temperature of the die during sealing should not exceed a safe level.

Solders and glasses are the common lid seal. Glasses are used as sealing materials mostly in commercial applications. They have the advantages of low cost, electrical insulating properties, and closely matched coefficients of thermal expansion with lid and case materials. The main disadvantages of glasses are their high sealing temperatures and their brittleness. The sealing temperatures of even low melting glasses range from 400 to 450°C. However, these comparatively low sealing temperatures of glasses result in relatively high coefficients of thermal expansion (8.5–10.0 X 10^{-6}/K), which can contribute to greater strains in the seal during temperature excursions. The composition of glasses can be varied to give a wide range of CTEs. Properties of soldering and brazing alloys for lid sealing are listed in Table 31.

The most commonly used solder is 60Sn-40Pb alloy, with traces of other elements to attain special properties. This solder, the tensile strength of which is 18.6–28.0 MN/m², is used for low-strength seals. Soldering, however, does not subject the electronic device to high temperatures during the sealing process, as the melting point for Sn60Pb40 solder is about 191°C and that of other commonly used solders varies from 117°C to 310°C. The coefficient of thermal expansion of solders used most frequently for lid sealing varies between 13 and 30 ppm/K. This does not provide a close match with commonly used lid/case metals, such as Kovar, which has a coefficient of thermal expansion of 5.1–5.87 ppm/K, potentially resulting in large thermomechanical stresses. Solders also have the disadvantage of poor corrosion resistance. Gold eutectic alloys, typically 80Au20Sn eutectic, can be used instead of soft solders when higher strength and greater corrosion resistance are required.

Table 29. Properties of package case materials (Pecht 1991)

Material	Tensile strength (MPa)	Compressive Strength (MPa)	Flexural strength (MPa)	Dielectric constant @ 1MHz	Resistivity (Ω-cm)	Thermal conductivity (W/m-°C)	CTE (ppm/°C)
BeO	230	-	250	7.0 – 8.9	10^{11}–10^{12}	150–300	6.8–7.5
SiC	17.24	490	440–460	2.0 – 4.2	$>10^{11}$	120–270	3.5–4.6
AlN	-	392–441	360–410	8.8	$>10^{11}$	170–260	4.3–4.7
Al_2O_3	124–207	1620–2758	290–414	8.8	$>10^{12}$	15.33	4.3–7.4
Mullite $(3Al_2O_3\ 2SiO_2)$	-	-	125–275	6.8	$>10^{12}$	5.0	4.2
Cu:W(10:90)	489.5	-	1062	-	6.6×10^{-8}	209.3	6.0
Kovar	552	-	-	-	50×10^{-8}	15.5–17	5.87
SiC-Al Composite	-	-	-	-	-	160	8.2

Table 30. Sealing glasses used for hermetic packaging

Glass type KC no.	Glass code number	Sealing temperature (°C)	Softening temperature (°C)	CTE (ppm/°C)	Thermal conductivity (W/cm-°C)	Dielectric constant @ 1MHz, 25°C	Volume resistivity (Ω-cm)
7583	1VP7583	400	385	7.8	0.0027	18.8	10^9
KC-1	LS-0110	450	387	5.25	0.0032	11.8	10^9
KC-1M	LS-0113	450	400	6.1	0.0034	35.0	$10^{9.5}$
KS-1175X1	XS-1175X1	430	375	6.16	0.0028	12.2	10^9
KC-402	T-1918F	430	350	6.7	0.0023	12.2	$10^{9.4}$
KC-405	LS-2001B	430	350	6.96	0.0022	14.0	10^9
KC-700	T-187HP	430	342	6.8	0.0028	12.5	$10^{9.4}$
KC-810	LC-3001	415	350	6.94	0.0029	12.2	$10^{9.6}$
KC-400	LS-0803	400	342	6.59	0.0037	35.2	10^9
KC-900	New Glass	400	340	6.46	0.0034	28.4	10^9

Table 31. Properties of soldering and brazing alloys for lid sealing

Alloy	Composition	Tensile strength (MPa)	Yield strength (MPa)	Shear strength (MPa)	CTE (ppm/°C)	Range Liquidus (°C)
Sn96	0.1Pb:3.6-4.4Ag:Sn	36.3-57.6	48.8	32.1	29.3	221
Sn63	62.5-63.5Sn:0.2-0.5Sb:0.015Ag:Pb	35.4-42.2	16.1	28.5	24.7	183
Sn62	62.5Sn:0.2-0.5Sb:1.75-2.2Ag:Pb	31.0-59	17.7	27.6-37.9	27.0	179
Sn60	59-61Sn:0.2-0.5Sb:0.015Ag:Pb	18.6-28.0	14.2	24.1-33	23.9	191
Sn50	49-51Sn:0.2-0.5Sb:0.015Ag:Pb	24.2	46.5	24.2		
SnIn50	50Sn:50In	11.8		11.2		117
AgIn90	10Ag:90In	11.3		11.0		141
PbIn25	75Pb:24In	37.5		24.2		226
PbIn5	95Pb:5In	29.8		22.2		
Sn10	9-11Sn:0.2Sb:1.7-2.4Ag:Pb19.7-24.3	19.7-24.3	13.9	19.3		290
Sn5	4.5-5.5Sn:0.5Sb:0.015Ag:Pb	23.2	13.3			312
Sb5	94Sn:0.2Pb:4-6Sb:0.015Sb	56.2	38.1		28.7	240
Ag1.5	0.75-1.25Sn:0.4Sb:2.3-2.7Ag:Pb	38.5	29.9	21.0		309
Ag2.5	0.25Sn:0.4Sb:2.3-2.7Ag:Pb	26.5	16.5	17.9		304
AuSn	20Sn:80Au	198		185	16.0	280
AuSi	97Au:3Si	255-304			10-12.9	
AuGe	88Au:12Ge	233		220	13.0	356
PbIn50	50Pb:50In	32.2		18.5	26.3	180

However, peak sealing temperatures of about 350°C are required, which are considerably higher than the temperatures needed for soft solders.

3.5   Leads

The main properties to be considered in the selection of materials for leads and leadframes are electrical conductivity, corrosion resistance (determined by the position of the material in the galvanic series), solderability, coefficient of thermal conductivity, yield strength, fatigue properties, and coefficient of thermal expansion. The lead material should be electrically conductive to serve as the electrical path for the signals. The lead material should be resistant to corrosion, which increases the electrical resistance of the leads, causing electrical failure, and can eventually result in mechanical fracture. Leads have to be soldered to the board, and hence should be wetted by the solder. Leads can serve as a path for dissipation of heat generated during device operation, so they require high thermal conductivity. Yield strength and ductility are needed for formability (the ability to accommodate sharp bends). Fatigue properties characterize the ability of the lead to withstand cyclic stresses typically induced under vibrational or temperature cycling loads.

Lead materials in current use include Cu-Fe, Cu-Cr, Cu-Ni-Si or Cu-Sn alloys. Special alloys like ASTM F30 or Alloy 42 (42Ni-58Fe) and ASTM F15 or Kovar (29Ni-17Co-54Fe) have gained wide acceptance because of their thermal expansion coefficients, which closely match those of ceramics, and their high formability. Properties of lead/lead frame alloys are listed in Table 32.

Alloy 42 and Kovar are commonly used for lead and leadframe fabrication in ceramic chip carriers. The coefficient of thermal expansion of Kovar is 5.1 to 5.8 ppm/°C, and that of Alloy 42 is 4.0–4.7 ppm/°C, in the temperature range 20–300°C. Thus, the coefficients of thermal expansion of both these materials match well with those of silicon which are 2.3 ppm/°C, and that of ceramic substrate (3.4 to 7.4 ppm/°C). Kovar and Alloy 42 also have a high fatigue strength. Alloy 42 has a fatigue strength of 620 MPa compared with only 380-550 MPa for most copper alloys.

Copper alloys possess the advantage of high electrical conductivity, which is critical because increased resistivity causes signal attenuation and the loss of switching speed. Electrical resistivity of the copper alloys in common use lies between 1.8–4.9 mW-cm compared with 49 mW-cm for Kovar and 70 mW-cm for Alloy 42. The thermal conductivity of Cu alloys is also greater than that for Kovar or Alloy 42. The thermal conductivity of CuW is 209 W/m-°C, compared with only 40 W/m-°C for Kovar and 12 W/m-°C for Alloy 42. High conductivity makes copper alloys particularly suitable for leadframes of plastic packages, where the leadframe constitutes the major heat dissipation path. Copper alloys also exhibit better solderability characteristics than Kovar and Alloy 42. Alloy 42 and Kovar typically have a nickel undercoat to enhance solderability and a gold overcoat to prevent surface oxidation during storage.

Table 32. Properties of lead/lead frame alloys

Alloy Group	Symbol	Nominal composition	CTE (ppm/°C)	Thermal conductivity (W/m·°C)	Electrical resistivity (μΩ-cm)	Yield bend fatigue strength* (MPa)
Cu-Fe	C19400	2.35Fe0.03P0.12Zn	17.4	260	2.54	475
	C19500	1.5Fe0.8Co 0.05P0.6Sn	16.9	200	3.44	-
	C19700	0.6Fe-0.2P 0.04Mg	-	320	2.16	450
	C19210	0.10Fe 0.034Mg	-	340	2.03	380
Cu-Zr	CCZ	0.55Cr0.25Zr	-	340	2.03	430
	EFTEC647	0.3Cr0.25Sn 0.2Zn	-	-	-	-
Cu-Ni-Si	C7025	3.0Ni0.65Si 0.15Mg	17.2	160	4.31	620
	KLF 125	3.2Ni0.7Si 1.25Sn0.3Zn	-	140	4.89	-
	C19010	1.0Ni0.2Si 0.03P	-	240	2.87	585
Cu-Sn	C50715	2.0Sn0.1Fe 0.03P	-	140	4.89	550
	C50710	2.0Sn0.2Ni 0.05P	17.8	120	5.75	450
Others	C15100	1.0Zr99.0Cu	17.6	380	1.81	380
	C15500	0.11Mg0.06P99.83Cu	-	344	1.99	-
Fe-Ni	ASTM F30 (Alloy 42)	42Ni58Fe	4.0-4.7	12	70	620
Fe-Ni-Co	ASTM F15 (Kovar)	29Ni17Co 54Fe	5.1-5.87	40	49	-

*Load bend fatigue strength of alloys capable of withstanding 4–5 cycles before failure in 0–90–0 degree cycles, which is above the three-cycles-to-failure minimum in MIL-STD-883. Values pertain to a 0.25 mm thick strip sheared to 0.45 mm width.

4   SECOND-LEVEL PACKAGING MATERIALS

At the second level of electronic packaging, components are mounted on a platform known as a printed wiring board (PWB) or printed circuit board (PCB). The board is either single-sided, double-sided, or multi-layered, depending on the number of components and the interconnection density. Components are interconnected to one another by the conductors embedded among the layers of the board or on its surface. Power and ground are provided to the components by point-to-point conductors in a single- or double-sided PWB or through planes embedded in the layers of a multi-layer board. A typical PWB serves an electronic system mechanically, electrically, and thermally. Mechanically, it provides support for the components and the conductors and a thermal conduction path for the heat dissipated by the components. Electrically, the PWB provides an insulator for the conductors. Because moisture absorption and chemical attack degrade material properties, the chemical properties also need to be considered.

Monolithic materials, such as ceramics and polymeric films, as well as fiber-sheet reinforced composite materials, such as glass-epoxy, are employed as PWB materials. Two types of wiring boards are commonly used: rigid printed wiring boards (PWBs) and flexible wiring boards (FWBs). Rigid PWBs are fabricated either from ceramics or from fiber-reinforced thermoset polymers. Compliant polymeric films are typically employed for FWBs. The wiring pattern is formed by subtractive, additive, or semi-additive processes. In the subtractive process, a board layer is clad with a conductor metal sheet that is selectively etched for the wiring pattern. An additive process involves no etching; the wiring pattern is obtained by electroless deposition. In the semi-additive process, a layer of copper is electroless-deposited on the entire unclad board and the wiring pattern is obtained using photo-imaging. Wiring traces are electroplated to the required thickness over the exposed copper. This is followed by stripping the photoresist and quick-etching the copper exposed by the stripping.

The common fiber and thermosetting polymeric resin for rigid PWBs are described in Sections 4.1 and 4.2, respectively. Common rigid PWB laminate materials, including ceramics, are discussed in Section 4.3, followed by a discussion on constraining cores in Section 4.4. Materials for flexible printed wiring boards, conductor metals in wiring boards, and conformal coatings are discussed in Sections 4.5 to 4.7.

4.1   Reinforcement Fiber Materials

Composite PWB materials are laminates made by bonding together several plies of fiber-reinforced polymeric resin. The desired effective properties of the laminate are obtained by off-setting the mechanical or thermomechanical weakness of the embedding resin by reinforcing fibers. Sheets of a fiber reinforcement employed in PWB laminates are in the form of either woven fabrics or paper. Woven fabrics impart bi-directional thermal stability to each layer by virtue of the orthogonal yarns. The choice of fabric weave is primarily governed by thickness and composite effective property requirements, rather than by impact toughness, specific strength, and specific stiffness, which are of primary interest in structural applications. Therefore, balanced woven-fabric reinforcements (the number of yarns and fibers per yarn) are the same per unit length along warp and fill directions, with relatively high resin-volume fractions popular in PWB applications.

Common fiber materials include inorganic glasses and organic polymers. Table 33 lists the properties of selected fiber materials.

Glasses are supercooled inorganic liquid materials. The glasses used as fiber reinforcements in the PWB industry include E-glass, S-glass, D-glass, and quartz. These fibers possess isotropic properties approximately as listed in Table 33. The chemical composition of these glasses are listed in Table 34.

E-glass fibers are the predominantly used reinforcement in PWB applications today. In fabric form E-glass fibers are readily available in various weave styles, as listed in Table 35. The fabrics are easy to handle and form a good bond with most embedding resins.

Table 33. Properties of common fiber materials
(Minges 1989, Du Pont 1988, Mumby 1989a)

Property	E-Glass	S-glass	D-glass	Quartz	Kevlar-49	e-PTFE
Specific gravity	2.54	2.49	2.16	2.2	1.44	2.2
Elastic modulus (GPa)						
Axial	72.4	86.2	51.7	68.9	124	8.9
Transverse	72.4	86.2	51.7	68.9	6.9	-
Axial tensile strength (GPa)	3.44	4.58	-	1.96	3.62	-
CTE (ppm/°C)						
Axial	5.5	2.6	3.0	0.54	-5.2	-
Transverse	5.5	2.6	3.0	0.54	41.4	-
Thermal conductivity (W/m-°C)						
Axial	1.0	3.0	-	1.0	0.25	0.31
Transverse	1.0	3.0	-	1.0	0.025	-
Moisture absorption (weight %)	0	0	0	0	0.5-2.0	-
Dielectric constant @ 23°C, 1 MHz	6.3	6.0	4.6	3.7	3.7	2.0
Loss tangent @ 23°C, 1MHz	0.0037	0.002	0.0015	0.0002	0.002	0.0002
Poisson's ratio	0.22	0.22	0.22	0.22	0.36	-

E-glass fibers possess good resistance to water, fair resistance to alkalis, and poor resistance to acids. These fibers exhibit very little deterioration in properties over time. The key disadvantage of E-glass fibers is their high dielectric constant, which limits their use in high-speed circuit applications.

S-glass fibers are being introduced as a replacement for E-glass fibers. Compared with E-glass, S-glass fibers have a 33% greater tensile strength, about a 20% higher tensile modulus, a reduced dielectric constant of 6.0, and approximately half the loss tangent (0.002). The specific gravity of S-glass fibers is comparable to E-glass fibers.

D-glass fibers were specifically developed with improved dielectric properties for high-speed electronic applications. The electrical properties of D-glass include a dielectric constant of 4.6 and a loss tangent of 0.0015, which are low compared with other glasses. Other physical properties of D-glass include a CTE of 3.0 ppm/C and specific gravity of 2.16. The modulus of elasticity of D-glass is, however, lower than E-glass by about 25 percent. In the U.S., D-glass fibers are not currently available.

Aramid is a family of polymeric fiber available from Du Pont De Nemours & Company known by the trade name Kevlar. The family includes Kevlar-49 which has been specially designed for PWB applications.

In woven form, aramid fabrics are available in weave styles 108, 120 and 177. Aramid fibers are transversely isotropic (refer to Table 33), with an orthotropy ratio (transverse to axial property) in the range of 0.1 to 10, depending on the property. Compared to E-glass fibers, Kevlar-49 fibers have almost twice the axial strength, approximately half the specific gravity, and better fatigue and vibration damping.

Table 34. Chemical weight percent composition of fiberglass
(Lubin 1986, Mumby 1989a)

Composition	E-glass	S-glass	D-glass	Quartz
SiO_2	54.5	65	74.5	99.7
Al_2O_3	14.5–15.2	25	0.3	trace
Fe_2O_3	0.05–0.40	0.21	trace	trace
CaO	17.2	0.01	0.5	trace
MgO	4.7	10.2	-	-
B_2O_3	8.0	0.01	-	-
TiO_2	0–0.8	-	-	-
BaO	-	0.2	-	-
K_2O	0–1.4	-	-	-
Na_2O+K_2O	Trace	-	2.5	trace
F_2	0–1	-	-	-

The tensile strength of Kevlar-49 fibers reduces appreciably above 200°C due to aging. The reported value of the dielectric constant is 3.7 and the loss tangent is 0.002 (Mumby 1989a). Compared to E-glass fibers, Kevlar-49 fibers are advantageous for high-speed applications.

The thermal conductivity and thermal dilation of aramid fibers are highly orthotropic. Thermal conductivity along the fiber axis is approximately 1.68 W/m°C and along the radial direction, 0.14 W/m°C. The CTE along the axis of the fibers is − 5.2 ppm/°C; in the radial direction it is 41.4 ppm/°C. During processing, substantial shifts among layers can occur due to a large mismatch in the CTEs of the resin and the fibers. The main disadvantage of aramid fibers is water and solvent absorption, which make processing difficult. Kevlar-49 fibers begin to decompose at 427°C.

Quartz fibers, drawn from fused silica, possess a CTE approximately one-tenth that of E-glass fibers, along with a good combination of electrical and mechanical properties. The tensile modulus is comparable to that of E-glass fibers, a much lower dielectric constant (3.7), a much lower loss tangent (0.0002), and a lower specific gravity (2.2). In most respects, quartz fibers behave like glass fibers and can be embedded in virtually any resin system currently being used. Two woven fabric styles are available: style 503 which yields about 5–6 mils per ply at 57–60% resin content by volume, and style 525, which yields about 3–4 mils at slightly higher resin content. The fabrics are much more expensive than E-glass fabrics. Quartz fibers are relatively very fragile because of the hardness of the fused silica. Drill bits wear out much faster when drilling through laminates reinforced with quartz fibers than they do with E-glass/epoxy laminates.

Polytetrafluoroethylene is available in unexpanded fiber form as TFE-fluorocarbon, marketed by Du Pont under the trade name of Teflon, and in expanded form as e-PTFE fiber from W.L. Gore and Associates.

In the expanding process, porosity is introduced into the fibers to enhance the electrical properties. Due to the expansion, e-PTFE fibers have higher tensile and compressive strengths and lower breaking elongation than their resin counterparts, while retaining most of the physical and chemical properties of resin. Resin tensile strength is rated at 5,000 psi, an order of magnitude lower than the fiber tensile strength of 75,000 to 100,000 psi. The dielectric constant of e-PTFE is 1.7, which is lower than the resin from which it is expanded, due to the introduced porosity. The effective dielectric constant of fibers rises when used in a composite, due to the embedding resin permeating the pores of the fibers. However, this does not affect the tensile strength of the fibers. E-PTFE has a very high elongation at a yield of 17.6 percent compared to Kevlar-49 (with 1.3 percent) and brittle E-glass and, therefore, laminates employing e-PTFE as resin are easily bendable. Thermal conductivity of e-PTFE is poor and is comparable to that of epoxy resins. Chemical resistance and electrical properties of e-PTFE fibers are the best of the fiber materials used today for PWBs.

Table 35 E-glass fabric weave styles.

Style	Count (End/5cm) Warp x Fill	Yarn (Tex/Metric) Warp x Fill	Nominal Thickness (mm)	Weight (g/m^2)
104	118x102	55.51x052.75x0	0.028	18.6
106	110x110	55.51x055.51x0	0.033	24.4
108	118x93	55.51x255.51x2	0.061	47.5
112	79x77	5111x25111x2	0.092	70.5
113	118x126	5111x255.51x2	0.087	81
116	118x114	5111x25111x2	0.102	105
119	106x98	5111x25111x2	0.091	91.8
1070	118x69	5111x055.51x0	0.046	34.2
1080	118x93	5111x05111x0	0.053	46.79
1116	118x114	5221x05221x0	0.089	104
1165	118x102	5111x29331x0	0.101	121
1180	118x98	5111x05111x0	N.A	48.9
1316	120x120	5221x05221x0	0.097	108
1675	79x63	6331x06331x0	0.101	96.3
2112	79x77	7221x07221x0	0.081	69.2
2113	118x110	7221x05111x0	0.079	77.3
2116	118x114	7221x07221x0	0.094	102
2116	106x98	7221x07221x0	0.086	90.2
2125	79x77	7221x09331x0	0.091	86.1
2313	118x126	7221x05111x0	0.084	80.5
2316	120x120	7221x07221x0	0.096	106
2319	1178x91	7221x07221x0	0.086	92.2
7627	87x59	9681x09681x0	0.165	199
7628	87x63	9681x09681x0	0.173	204.4
7629	87x67	9681x09681x0	0.180	213
7637	87x43	9681x091361x0	0.224	228
7642	87x39	9681x091361x0(TEX)	0.254	228
7650	87x45	9681x09991x0	0.190	208
7652	63x63	9991x09991x0	0.220	252
7660	59x59	9681x09681x0	0.150	160.4

Note: Above values are industry targets. Al styles above are plain weave. Constructions with identical yarn counts, but having yarns with different filament diameters, shall be designed by changing the first digit of the style number to the next lowest available number, for example, 2116 to 1116.

However, dimensional stability is a problem with laminates reinforced with e-PTFE fibers, due to fibers being weak in compression, leading to unconstrained shrinkage of resin after etching copper from clad base material. Therefore, 10 to 12 percent by volume of a co-fiber, such as E-glass or quartz, is used to provide the required compressive strength.

4.2 Resins

Polymeric resins are the matrix material used in most electronic reinforced laminates. Almost all rigid printed wiring board laminates employ thermoset polymer resins. Compared with the thermoplastic resin laminates, thermosetting resin laminates exhibit improved temperature performance and chemical stability. However, thermosetting resin laminates cannot be molded because polymerization of thermoset resins leads to rigid cross linking among molecular chains which is irreversible. Further, at temperatures beyond T_g, they degrade.

The polymerization process is also called curing, which is achieved primarily in three stages. The liquid form is known as stage A, followed by a partially cured form, stage B, and a fully cured form, stage C. Curing of resin in stages is of special advantage in the construction of the multi-layer structure of a laminate. To obtain the required thermal stability and stiffness of each layer during lamination prepregs, stage B of resin impregnated fiber sheets are used.

Thermoset resins available for fabrication of PWBs include epoxies, polyimides, and tetra-fluoro-ethylene. Typical properties of important resins are listed in Table 36.

Epoxy resins are the most common resin types used for PWB fabrication. The properties that make epoxies popular are excellent adhesive bond strength with metals, good electrical insulation, good mechanical strength, low shrinkage, high chemical resistance, and excellent processability due to controlled resin viscosity and exothermic reaction during curing. Moreover, epoxy is readily modified for tailoring of properties through the use of different curing agents.

Epoxy is a reaction product of almost exclusively diglycidyl-ether of bisphenol-A (DGEBA)-based resin and a curing agent, such as an amine. Due to rotational freedom of the hydroxyl group in the molecular structure of DGEBA, epoxies based on this resin exhibit a large variation in dielectric constant and dissipation factor (Mumby 1989c). Bromination of DGEBA renders the epoxy flame retardant. FR-4 grade epoxy is based on brominated DGEBA and the curing agent dicyandiamide (DICY). The B-stage of FR-4 grade epoxy has excellent long-term stability. However, the glass transition temperature of FR-4 grade epoxy is about 120 °C, which is lower than solder temperatures and further creates problems due to the high CTE of epoxy beyond T_g.

Unreacted dicyandiamide in the cured laminate adversely affects the glass transition temperature, leads to increased water absorption and degrades electrical properties of the epoxy. Quatrex, introduced by the Dow Chemical Company, is one of the many non-DICY epoxies available today. Di-functional and multi-functional epoxies are modified epoxies employed in applications where a higher glass transition temperature, improved thermal stability, and better chemical resistance are required.

Table 36 Physical properties of typical resin systems (Harper 1991, Mumby 1989a)

Resin System	Temperature Tg (°C)	Elastic Modulus (GPa)	Poisson's Ratio	CTE (ppm/°C)	Thermal Conductivity (W/m°C)	Dielectric Constant @ 23°C, 1MHz	Loss Tangent @ 23°C, 1MHz
Epoxy	120-140	3.45	0.37	69	0.19	3.6	0.032
Polyimide	240-300	2.8	0.33	50	0.18	3.2	0.020
Cyanate Ester	260	2.6	-	55	0.20	3.1	0.005
PTFE	-	0.35	0.46	99	0.19	2.0	0.0002
BT	250	-	-	-	0.20	3.1	0.003

The modified epoxies have a T_g in the range of 140 to 180 EC. The increased cross-link density in the modified epoxies increases the brittleness and reduces ease of processing.

Typical mechanical and electrical properties of some epoxies, along with their resin and curing agents, are listed in Table 37. The effect of the curing agent on deflection temperature is significant.

Polyimides used for rigid printed wiring boards are the reaction product of aromatic diamine and bismalemide. The use of bismalemide is preferred over aromatic dianhidride or tetracarboxylic acid to avoid condensation reactions. Polyimides obtained from a condensation reaction rapidly increase in viscosity, leading to voids in laminates. However, films have been successfully manufactured for flexible wiring boards using the condensation reaction.

Polyimides exhibit good thermal performance at elevated temperatures and properties comparable to epoxies at lower temperatures. The glass transition temperature of polyimide resins is in the range of 260 to 300 °C. They have excellent resistance to short-duration exposure to high temperatures, as indicated by a time-to-blister at 290 °C of over 1200 seconds. Drill smear does not occur in multi-layer boards made with polyimide resin. Heat applied during solder rework and during lamination tends to increase adherence of the metal conductors.

Disadvantages of polyimides include high cost, a low flammability rating of UL-94 V-1, and higher moisture absorption. Absorbed moisture degrades polyimide's electrical properties, though under anhydrous conditions, polyimides have lower dielectric constants and loss tangents than FR-4 epoxy. Polyimide boards are brittle and cannot be easily punched. Extended exposure to high temperature causes darkening of color, which is cosmetic and does not affect the electrical and physical characteristics of the resin.

Cyanate ester resins are formed by the cyclotrimerization reaction of aryl dicyanates (Mumby 1989a). Advantages of cyanate ester resins include their low dielectric constant and loss factor, which vary little with temperature and electromagnetic frequency compared with other thermosetting resins. Further, the physical properties of the cyanate ester resins are tailorable by the proper choice of the aryl dicyanate molecular structure. Cyanate ester resins are one of the most promising resins for high-speed applications. They are available from many manufacturers such as Dow Chemical Company and Hi-Tek Polymers, Inc.

Cyanate ester resins have glass transition temperatures greater than 180 °C, dielectric constants in the range of 2.8 to 3.0, and a loss tangent of approximately 0.004. However they are much more expensive than the standard FR-4 grade epoxy. The major disadvantage of cyanate ester resins is their high moisture absorption; recently, this problem of cyanate esters has been attributed to carbonate impurities in the resin and has been controlled by purification. Bromination of a cyanate ester resin yields a flame-retardant resin.

Table 37. Mechanical and electrical properties of selected epoxy resins (Clayton 1989)

Property	Anhydride	Aliphatic	Aromatic amine	Catalytic	High-temperature anhydride	Epoxy Novolac anhydride	Di-anhydride
Resin / Curing agent	DGEBA/ HHPA	DGEBA/ DETA	DGEBA/ MPDA	DGEBA/ BF3MEA	DGEBA/ NMA	NOVOLAC/ NMA	DGEBA/ PMDA
Heat deflection temperature(°C)	130	125	150	174	170	195	280
Tensile strength (MPa), @ 23°C	72	75	85	43	75	66	22
Tensile modulus (GPa), @ 23°C	2.80	2.87	3.30	2.70	3.40	2.94	2.70
Tensile strength (MPa), @ 100°C	37	32	45	29	46		14[b]
Tensile modulus (GPa), @ 100°C	2.10	1.80	2.20	1.90	1.40		
Flexural strength (MPa), @ 23°C	126	103	131	112	112	147	59
Flexural modulus (GPa), @ 23°C	3.22	2.48	2.80		4.80	3.69	3.61
Compressive strength (MPa), @ 23°C	111	224	234		116	159	254
Compressive modulus (GPa), @ 23°C	5.09	1.86	2.31		0.73	2.22	2.41
Volumetric resistivity (Ω-cm), @ 23°C	4e14	2e16	1e16		2e14	1e16	
Dielectric constant, @ 23°C and 1 MHz	3.2	3.33	3.85	3.20	2.99	3.20	3.34
Dissipation factor, @ 23°C and 1 MHz	0.013	0.034	0.038	0.024	0.021	0.016	0.022

NB.: HHPA, hexahydrophthalic anhydride; DETA, diethylenetriamine; BF3 MEA, boron trifloride monoethylamine complex; NMA, nadic methyl anhydride; PMDA, pyromellitic dianhydride; BDMA, benzyldimethylamine b: Measured @ 150°C.

BT resins are blends of a bismaleimide and a triazine (cyanate ester resin) supplied by Mitsubishi Gas Company (Morio et al. 1985). The major advantages of the BT resins are high glass transition temperatures in the 180 to 190 EC range, retention of copper bond strength at elevated temperatures, high ductility, low dielectric constant, low loss tangent, and retained insulation resistance after moisture absorption. Different BT resins are obtained by varying the relative proportions of the two components. The T_g of BT resins increases with the increased proportion of bismaleimide, but high proportions of bismaleimide may degrade other physical properties of the resin. BT resins exhibit a good affinity to many thermosetting resins, such as epoxy and thermoplastic resins, and are easily modified.

Phenolic resins are condensation products of phenols and formaldehyde. The major advantages of phenolic resins include low cost, ease of processing, and excellent chemical and heat resistance. Phenolics used in lamination are resole resins, which are brittle due to their high degree of cross-linking in the polymer. Novalacs are sometimes added to resole resin to reduce their brittleness and improve resistance to water. The major disadvantages of phenolic resins include low insulation resistance, high water absorption, and brittleness. Generally, low-cost, paper-based reinforcements are used with phenolic resins to produce XXXP and FR-2 NEMA grade laminates (refer to Table 38).

PTFE or poly-tetra-fluoro-ethylene resin is the polymer resulting from the polymerization of tetra-fluoro-ethylene. The major advantage of PTFE resins is in high-speed applications, due to their low dielectric constants and loss tangents. The resin is produced by preforming and sintering of PTFE particles obtained from polymerization, and is usually carried out in aqueous emulsion or suspension. The physical properties of PTFE resin are almost identical to e-PTFE or Teflon fibers, except for the mechanical properties. Compared with epoxies, PTFE resins have some manufacturing and processing problems, such as layer-to-layer adhesion and electroless plating. Moreover, they are expensive.

4.3   Laminates

Most of the laminates used for rigid PWBs are classified by the National Electrical Manufacturers Association (NEMA) based on the combination of properties that determine suitability of a laminate for a particular end use. NEMA grades of widely used unclad laminates are listed in Table 38. Although NEMA specifications provide a basic guide to laminate properties based on laminate constituent resin and fiber material, other variables, such as the fiber volume fraction and the reinforcement micro-architecture, are also important determinants of a laminate's effective properties. Typical properties of selected PWB laminate materials are covered in Table 39.

Fiber reinforcements make laminate's effective properties orthotropic (Guiles 1990, Agarwal et al. 1991a,b), due to the inherent preferred fiber orientation. Balanced weave styles, with similar warp and fill yarn geometries, are popular in fabrics for printed wiring board reinforcement. These laminates exhibit one set of properties along the in-plane material principal directions and different properties along the out-of-plane direction. Bi-directional reinforcement using woven-fabric reinforced laminate couple the in-plane and out-of-plane properties. The coupling is a consequence of the volumetric nature of certain properties, such as the CTE and swelling due to water absorption as well as the directional transport by the preferentially oriented fibers.

E-glass/epoxy laminates are by far the most widely employed circuit board materials. The fire-retardant E-glass/epoxy laminate is specified by NEMA as grade FR-4, which is similar to NEMA grade G-10, but flame retardant. The glass transition temperature of FR-4 laminate ranges between 120 and 135 °C, which is acceptable for most commercial PWB applications. The advantages of E-glass/epoxy laminates are ease of processing, punchability, machinability, low cost, and availability. E-glass/epoxy laminates provide in-plane thermal expansion closely matched to inner-layer copper (16 ppm/°C), with fiber volume fraction in the range of 0.28 to 0.32 (Guiles 1990). This can enhance plated through-hole reliability (Bhandarkar 1991, Bhandarkar et al. 1991).

Typical values of effective orthotropic properties of FR-4 laminates (E-glass cloth-reinforced epoxy) are covered in Table 39. The resin volume fraction in a laminate can vary in the range of 40 to 75 percent, depending on the reinforcement weave style and processing conditions, such as temperature and pressure. The effect of resin volume fraction on laminate properties can be significant. Figure 2 shows the variation in laminate out-of-plane dielectric constant and loss tangent as a function of resin volume fraction for FR-4 laminates (Mumby and Yuan 1989b).

Effective orthotropic properties have been analytically obtained by Dasgupta et al. (1990) and Agarwal et al. (1991). Figure 3 shows laminate CTE, modulus of elasticity, shear modulus and Poisson's ratios as functions of the fiber volume fraction for E-glass/epoxy laminates (Bhandarkar 1991). The smallest periodic repeat volume of the laminate with appropriate boundary conditions has been exploited in obtaining these values. The properties of the constituent epoxy and E-glass fiber used in the simulation are listed in Table 40. These numerical simulation studies give a quantitative coupling among the in-plane and out-of-plane properties, and the intensity of dependence on resin and fiber properties. For example, Figures 3a and 3b show the lowering of in-plane and out-of-plane CTEs with reduced resin fraction. The in-plane CTE is consistently lower and the out-of-plane CTE higher than the epoxy CTE. Following the unit-cell approach of Dasgupta et al. (1990), thermal conductivity (Agarwal et al. 1991a, Dasgupta et al. 1991), dielectric constant, and loss factors (Agarwal and Dasgupta et al. 1993) have been obtained as a function of fiber volume fraction.

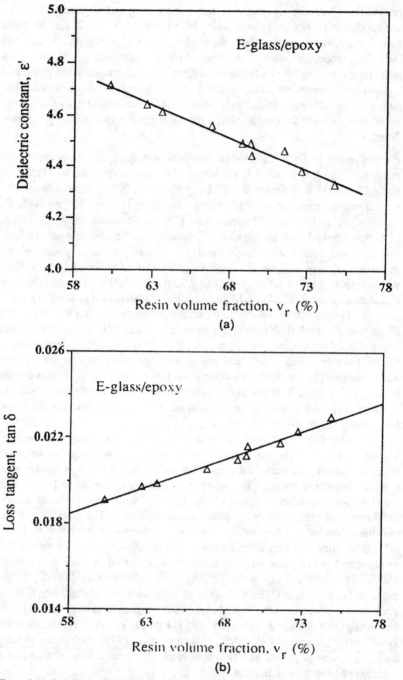

Figure 2: Variation in laminate out-of-plane dielectric constant and loss tangent as a function of resin volume fraction for FR-4 laminates at 1 MHz (Mumby and Yuan, 1989).

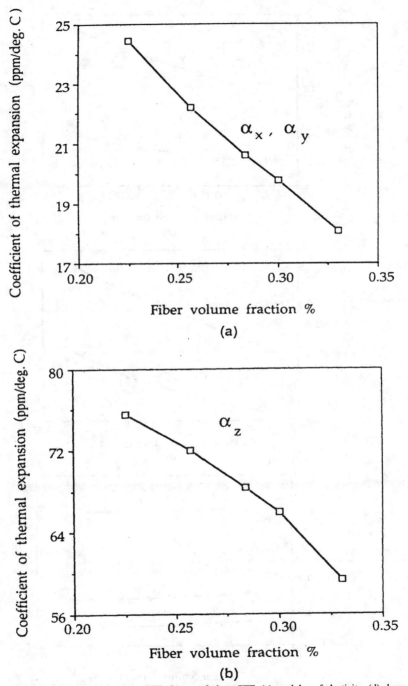

Figure 3: Laminate (a) in-plane CTE, (b) out-of-plane CTE, (c) modulus of elasticity, (d) shear modulus, (e) Poisson's ratio as functions of fiber volume fraction for E-glass/epoxy laminates (Bhandarkar, 1991).

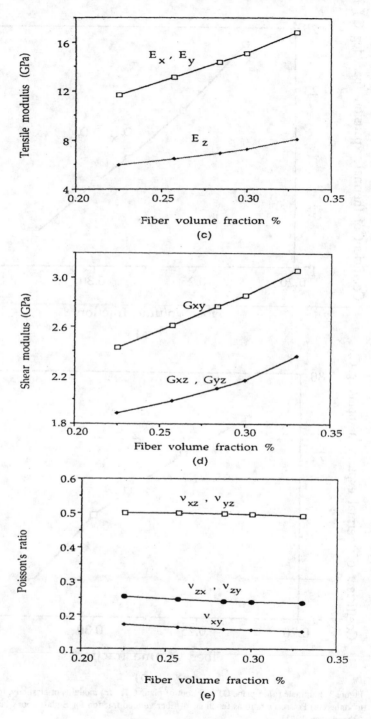

Figure 3 (continued).

Table 38. NEMA copper-clad laminates

NEMA Grade	Reinforcement	Resin	Description	Use
X	Paper	Phenolic	Good mechanical properties but low resistance to moisture	Used in inexpensive, low-cost consumer items
XX	Paper	Phenolic	Better moisture resistance than grade X	
XXP	Paper	Phenolic	Hot punch grade	
XXXPC	Paper	Phenolic	Room temperature punch grade	
CEM1	Paper and glass	Epoxy	Punchable and low flammability rating	Low-cost substitute for FR-4
G-10	Glass cloth	Epoxy	Excellent mechanical and good dielectric properties even under high humidity	Used in electrical applications but flammability is not a prime concern
FR-2	Paper	Phenolic	Equivalent to XXP and XXXPC but flame retardant	Used in inexpensive consumer items
FR-4	Glass cloth	Epoxy	Similar to G-10 but flame retardant	Used in computer applications
FR-5	Glass cloth	Epoxy	Similar to FR-4 but thermally more stable	Used in military applications
FR-6	Glass mat	Polyester	Random fiber and flame resistant	Used in applications where high impact toughness and low capacitance are desired
GI	Glass cloth	Polyamide	Thermally more stable than GF	Used in military applications
GT	Glass cloth	PTFE	Opaque brown laminate, with low dielectric constant and dissipation factor	For high-frequency applications
GX	Glass cloth	PTFE or polystrene polyethylene	Same as GT, but controlled electrical properties	For microwave applications

Table 39. Properties of selected PWB laminate materials
(Pecht 1991, Gray 1989, Mumby 1989a)

Laminate	Tg (°C)	CTE below Tg (ppm/°C)		Water Uptake MIL-P-13949F	Dielectric constant (@ 1MHz)	Dissipation factor (@1 MHz)	Tensile strength (MPa)	Modulus of elasticity (GPa)	Thermal conductivity (W/m°C)
		x,y	z		z	z	x, y	x, y	z
E-glass/ epoxy	120	12-16	60–80	10	4.7	0.021	276	17.4	0.35
E-glass/ polyimide	220–300	11-14	60–80	25	4.5	0.018	345	19.6	0.35
E-glass/ PTFE	75	24	261	-	2.3	0.006	68–103	1.0	0.26
Quartz/ polyimide	260	6–12	34	25	3.6	0.010	-	27.6	0.13
Quartz/ Quartrex	185	-	62	-	3.5	-	-	18.6	-
Kevlar-49/ Quatrex	185	3–8	105	10	3.7	0.030	-	22–28	0.16
Kevlar-49/ polyimide	180–200	3–8	83	25	3.6	0.008	-	20–27	0.12

Table 40. Properties of E-glass fiber and epoxy (Dasgupta et al. 1990)

E-glass fiber	Epoxy
E = 72.4 GPa	E = 3.45 GPa
v = 0.22	v = 0.37
CTE = 5.4 ppm/°C	CTE = 69 ppm/°C

E-glass/BT resin laminates offer manufacturing advantages identical to those of FR-4 laminates and sustained insulation resistance upon moisture absorption (Morio 1985). BT resin-based laminates exhibit effective properties comparable to the laminates based on epoxy resins, but with higher glass transition temperatures.

E-glass/polyimide, boosted by their ability to meet military requirements for improved reliability, in-service thermal performance, and field reparability, have evolved in the high-performance laminate market. An E-glass/polyimide laminate does not have a significantly better CTE than E-glass/epoxy below or beyond its T_g. However, the T_g of E-glass/polyimide is about 260°C, significantly higher than 120°C of E-glass/epoxy. Therefore, out-of-plane expansion and resin softening do not occur with E-glass/polyimide at solder reflow temperatures around 220°C. This avoids pad lifting and PTH damage during processing and enhances board reparability.

Quartz/polyimide laminates have CTEs in the range 7–9 ppm/°C, close to the CTE of ceramic chip carriers, making it suitable material for surface mountings. Due to the high T_g the plated through-hole reliability is good even at elevated temperatures. The dielectric constant of 3.5 to 3.8 in the 1 kHz to 1 GHz range and low-loss tangent of 0.008 make these laminates attractive for microwave polarizers or random applications. The disadvantage of quartz/polyimide laminates is the higher cost, four to five times that of E-glass/epoxy laminates.

Aramid/epoxy laminates are known for their in-plane CTEs, which are similar to those of ceramics. Aramid cloth-style 108 reinforced laminate exhibits in-plane CTEs in the range of 8 to 10 ppm/°C, whereas style 120 reinforced laminates exhibit 7 to 10 ppm/°C. However, the out-of-plane CTE of aramid/epoxy laminates is in the range of 90–115 ppm/°C below the T_g and in the range 280–340 ppm/°C above the T_g (Table 41). Compared with E-glass/epoxy laminates, the out-of-plane CTEs of aramid/epoxy laminates are higher. Other concerns in aramid/epoxy laminates are delamination and partially cut and torn fibers in the laminate during hole drilling. An etch back with a plasma etching process before plating is required for removal of partially cut fibers.

Aramid/polyimide laminates exhibit an in-plane CTE in the range of 6 to 7 ppm/°C, which matches that of ceramic components. However, due to the high transverse CTE of the fibers, brittle polyimide resin, and weave characteristics, these laminates exhibit micro-cracking of polyimide resin at the fiber-yarn cross-overs. The problem has been addressed through the use of Kevlar paper instead of cloth (Trace Labs 1986).

E-glass/PTFE laminates possess excellent electrical properties, chemical resistance, and thermal stability. The dielectric constant is approximately half of the FR-4 laminate and uniform over a wide bandwidth of the electromagnetic spectrum up to 15 GHz. The dissipation factor is two orders of magnitude lower than standard FR-4 laminate.

The glass transition temperature is 327°C. However, use of these laminates is not recommended above 120°C, due to the degradation of the copper-foil bond strength at prolonged temperatures above 120°C. Although E-glass/PTFE laminate itself can withstand higher temperatures, the copper to resin interface oxidizes and degrades. Special copper treatment is available to enhance bond strength by reducing the effects of oxidation.

**Table 41. Typical properties of woven Kevlar-49/Quatrex laminate
(COR-LAM,E4K) (Du Pont 1986)**

Property	Test Method	Typical Value	
Peel strength (kN/m)	IPC-2.4.8	0.53	
Glass transition temperature, Tg (°C)	IPC-2.4.25 (DSC)	180	
CTE (ppm/°C)	IPC-2.4.24 (TMA)	weave-108	weave-120
x, y (in-plane)		8–10	7–10
z (out-of-plane) below Tg		90–115	90–110
z (out-of-plane) above Tg		280–340	280–320
Thermal conductivity (W/m-°C)	ASTM-C-177	0.16	
Dielectric constant @ 1 MHz	IPC-2.5.5.3 ASTM-D-150	3.9	
Loss tangent @ 1 MHz	IPC-2.5.5.3 ASTM-D-150	0.013	
Electric strength (kV/mm)	IPC-2.5.6.2 ASTM-D-149	>100	
Volume resistivity (MΩ-cm)	IPC-2.5.17.1	$>1.0 \times 10^7$	
After moisture	ASTM-D-257		
At elevated temperature		$>1.0 \times 10^6$	
Surface resistivity (megohm/sq.)	IPC-2.5.17.1	$>1.0 \times 10^5$	
After moisture	ASTM-D-257		
At elevated temp.		$>1.0 \times 10^6$	
Arc resistance (sec.)	IPC-2.5.1 ASTM-D-149	>122	
Moisture absorption (wt %)	IPC-2.6.2.1	1.1	
Flammability	UL-94	V-O	
Solvent absorption (% max.)	IPC-2.3.4.2	3.0	

The cost of E-glass/PTFE laminates is twenty to thirty times more than FR-4 laminates. However, few other materials are capable of meeting the exacting requirements of high-frequency microwave applications.

Ceramic laminates offer the advantage of being able to sustain a harsher environment compared with organic laminates. The ceramic laminates are rigid (with a modulus of elasticity three to ten times that of organic boards), have high flexural strength (17–500 MPa), and are unaffected by most process chemicals and humidity. Their thermal stability is excellent, with a coefficient of thermal expansion in the 0 to 8 ppm/°C range. With ceramic components, a good CTE match enhances component-to-board interconnection reliability under cyclic thermal loads. Ceramics, unlike organic laminates, exhibit good thermal conductivity. The disadvantages of ceramic laminates include relatively high cost, high dielectric constant, and high density. Table 42 covers typical properties of ceramic substrate materials.

4.4   Constraining Cores

Constraining cores are generally high modulus sheets with low in-plane and out-of-plane CTEs, embedded in a laminate(s) to further tailor the "total" in-plane CTE. In addition, constraining cores are often tailored to provide high in-plane and out-of-plane thermal conductivity and low weight. The constraining cores are placed symmetrically in a laminate(s) to avoid warping.

Materials used for constraining cores include Inver and molybdenum. Often, in applications requiring light weight, a metal matrix composite such as graphite reinforced copper is employed. Constraining cores are often copper-clad to allow their use as power/ground planes. Cladded copper enhances the thermal conductivity of the core and can be laminated by conventional processes. The thickness of the copper foil is chosen depending on the desired thermal conductivity of the laminate. The out-of-plane thermal conductivity of the laminate is, however, dependent on the out-of-plane thermal conductivity of the metal core. Invar, an alloy of iron and nickel, has poor thermal conductivity relative to molybdenum and copper. Therefore, a copper-invar-copper (CIC) core has poorer out-of-plane as well as in-plane thermal conductivity than a sheet of copper of the same thickness. Copper-moly-copper (Cu-Moly-Cu) has an advantage over CIC, due to the good thermal conductivity of molybdenum; the core constructed with it has better in-plane as well as out-of-plane thermal conductivity. Recently copper-graphite has been developed for use as a constraining core, with the advantage of an improved specific modulus and strength (i.e., modulus and strength-to-density ratio). Table 43 lists typical properties of constraining cores.

4.5   Flexible Printed Board Materials

A flexible printed wiring board (FWB) is similar to a printed wiring board, but the FWB can be bent or folded. The FWB is used in applications where low weight and volume are required, where vibration is likely, and where the controlled impedance of long interconnection wiring is required. A printed wiring board in which a flex circuit constitutes some part of the circuit is termed a rigid-flex board. A FWB is comprised of a flexible substrate material, metal cladding, adhesive and, usually, a protective coating. In the following sections, common materials for these constituents and their typical properties are discussed.

Flexible films. Flexible films used as substrates for FWBs include polyimides, polyesters, nylons, fluorocarbons, and unwoven fiber-reinforced composites. Polyimide films such as Du Pont's Kapton Type H and polyester films such as Du Pont's Mylar Type A are the most widely used. Table 44 lists typical properties of some common films.

Polyimide films are available in electrical grade, with Kapton the most common. Kapton films possess excellent thermal performance and physical properties. The electrical properties of Kapton are good, as in other polyimides. The dielectric constant remains constant over a wide frequency band and the dissipation factor is low. The film is heat resistant and can be used in a broad temperature range of $-269^{\circ}C$ to $400^{\circ}C$. The disadvantage of Kapton is its relatively high moisture absorption. The absorbed moisture can degrade Kapton's electrical properties and lead to blistering and delamination of multi-layer FWBs.

Polyester films used in electronic applications are processed by melt extrusion followed by hot rolling and stretching. The films are partially crystalline, unlike amorphous polymers. The deformation-induced orientation of molecular chains imparts high strength and toughness to the film. Polyester films are available in many forms such as Du Pont's Mylar Type A. Advantages of Mylar include its resistance to water absorption, resistance to chemical attack, and low cost. The limitation of Mylar is that it is stable only up to 150 EC. Therefore, it is used where soldering is not required. Du Pont's fluorocarbon film, Teflon, has excellent electrical properties but requires special processing, which makes it very expensive for FWB applications.

Composite materials used for FWB include dacron-epoxy and Nomex by Du Pont. Dacron-epoxy consists of epoxy reinforced with a non-woven polyester fiber mat. Dacron-epoxy has a high fatigue endurance among the flexible films in use today. The film is solderable and flame-resistant, and its cost is comparable to Kapton. Coated aramid-fiber paper composite Nomex is hygroscopic, absorbing up to 13% moisture by weight, which adversely affects its dimensional stability due to swelling.

Flexible film substrates are available as single- and double-sided clad laminates. The metal claddings employed include copper foil, beryllium copper, aluminum, and Inconel. Rolled, annealed copper foil is the most widely used cladding material for FWBs.

Table 42 Ceramic printed wiring board substrate properties

Material	Max. use temperature (°C)	CTE (ppm/°C)	Thermal conductivity (W/m·°C)	Dielectric constant @ 1MHz	Loss tangent @ 1MHz	Tensile strength (MPa)	Flexural strength (GPa)	Modulus of elasticity (GPa)	Poisson's ratio
Aluminum nitride	1800	4.1	140–220	8.8	<0.001	193	400–500	275	0.25
Silicon carbide	1900	3.8	70	40	0.05	179	450	406	0.15
BeO	1700	7.2	260	6.7	0.0004	93	170–280	262	0.34
Boron nitride	2000	0.0	60	4.11	0.0045	45	-	43	0.23
Al$_2$O$_3$, 99%	1500	6.5	25	9.9	<0.0004	206	200–300	275	0.22
Al$_2$O$_3$, 96%	1500	7.1	20	8.9–10.2	0.001	172	200–300	275	0.22

Table 43. **Properties of constraining cores**

Constraining core	Density (g/cm^3)	CTE (ppm/°C) x, y	z	Therm cond (W/m-°C) x, y	z	Modulus of elasticity (GPa) x, y	Tensile strength (MPa) x, y
20Cu/60Invar/20Cu	8.45	6.02	7.7	164	22	135	310–412
12.5Cu/75Invar/12.5Cu	8.31	3.69	5.3	110	64	140	380–472
25Cu/50Mo/25Cu	9.56	7.9	-	268	-	220	427
13Cu/74Mo/13Cu	9.89	6.5	8.3	208	159	269	600

Rolled copper has higher fatigue ductility and is more flexible than electrodeposited copper. The electrodeposited copper used in FWBs is a high-ductility grade that exhibits high elongation.

Adhesives. Commonly used adhesives in FWBs include acrylics, epoxies, polyesters, fluorocarbons, and polyimides. A comparison of the adhesives is given in Table 45. Typical properties of adhesive acrylics, epoxies, and polyesters are listed in Table 46. Low cost, adhesiveless FWBs are produced by screen printing circuits directly onto flexible film using a low temperature curable conductive ink.

Acrylics are flexible, high-temperature adhesives with good electrical properties. Acrylics have the best resistance to short high-temperature exposure, and thus can be easily hand- or wave-soldered. Adhesive epoxies derive their flexibility from reduced crosslinking among polymer chains.

Epoxies exhibit thermal stability, good electrical properties, and chemical resistance. Fluorocarbons are high-temperature adhesives with excellent electrical properties, good chemical resistance, and good flexibility. However, they have poor dimensional stability. Polyimides have the highest thermal stability among flex-board adhesives and can withstand temperatures up to 370°C. However, they have low inter-layer bond strength and are relatively less flexible than other adhesives. Polyester adhesives are low-temperature thermoplastic polymers. They have low cost, excellent flexibility, good electrical properties, and good chemical resistance.

Copper is processed, primarily, in three ways to obtain properties to suit the application. The three processes are electrolysis, rolling, and chemical reduction. Depending on the manufacturing process, the metal is known as electrodeposited, rolled, or electroless copper. The electrodeposited and rolled (wrought) copper are available as foils.

Table 44. Typical properties of flex substrate materials (Du Pont 1990)

Property	Test method	Polyimide (Du Pont's Kapton)	Polyester (Du Pont's Mylar)	Fluoropolymers
Tensile strength (MPa)	ASTM D 882	230	170	20–28
Elongation (%)	ASTM D 882	72	150	300
Tear strength (MD) propagating, (g)	ASTM D 1922-67	7	1.7–2.5	12
Specific gravity	ASTM D 1505	1.42	1.39	2.2
Thickness range (mils)	-	0.3–5	1.2–7.5	1–7
Dimensional stability maximum (%)	IPC-TM-650 Meth-2.2.4-A	0.17–1.25	1.5	0.3
Max. use temperature (°C)		240	150	210
Flammability (Oxygen index)	IPC-TM-650 Meth-2.3.8	30	22	30
CTE (ppm/°C)	-	18–50	31	10–13
Dielectric strength (volt/mil) 1 mil-film 23°C	ASTM D 149	7700	7500	5000
Dielectric constant @ 10 Hz @ 1MHz	ASTM D 150	3.5 / 3.4	3.25 / 3.0	2.0 / 2.0
Loss tangent @ 10 Hz @ 1MHz	ASTM D 150	0.0025 / 0.01	0.005 / 0.016	0.0002 / 0.0003
Volume resistivity (Ω-cm)	ASTM D-257-78	1.5×10^{17}	-	-
Surface resistivity (MΩ/sq.)	IPC-TM-650 Meth 2.5.17	10	-	10
Chemical resistance to :				
Acids	ASTM D 543	Good	Good	Good
Alkalis	ASTM D 543	Poor	Poor	Good
Grease and oil	ASTM D 722	Good	Good	Good
Organic solvents	ASTM D 543	Poor	Good	Good
Fungus resistance	IPC-TM-650 Meth 2.6.1	Non-nutrient	Non-nutrient	Non-nutrient
Moisture absorption (wt. %)	IPC-TM-650 Meth 2.6.2	3	0.8	0.1

Table 45. Comparison of adhesives used in flexible wiring boards (Shepler 1989)

Property	Polyester	Acrylic	Modified epoxy	Polyimide	Fluoro-carbon	Butyral phenolic
Temperature resistance	Fair	Very good	Good	Excellent	Very Good	Good
Chemical resistance	Good	Good	Fair	Very Good	Very Good	Good
Electrical properties	Excellent	Good	Good/Excellent	Good	Good	Good
Adhesion	Excellent	Excellent	Excellent	Very Good	Very Good	Good
Flexibility	Excellent	Good	Fair	Fair	Excellent	Good
Moisture absorption	Fair	Poor	Good	Poor	Excellent	Fair

Electrodeposited copper: Electrodeposited (ED) copper foil is the standard copper used in rigid laminates. Sheets of ED copper are electrolytically deposited from the copper anode onto a rotating steel drum cathode from which it is subsequently stripped. The side against the drum is smooth and shiny, while the other side has a grainy matte finish. The copper grains thus formed are elongated perpendicular to the surface of the foil (radial to the drum) providing "teeth" on the copper foil for adhesion with the prepreg resin. The adhesion is improved by oxide treatment or nickel flash. While the elongated grains are good for adhesion, they are not good for tensile strength and resistance to fatigue. To improve ductility and etchability the electrodeposited copper is stress relieved or annealed. Electrodeposited copper foil can have pin holes which can lead to resin bleeding during lamination. Some properties of copper are given in Tables 48-50.

Rolled copper is produced by successive reduction in thickness of a copper strip through rolling. The rolled copper is smooth on both sides and has grains elongated along the rolling direction. The grain structure provides fatigue endurance and elongation before rupture. However, the peel strength of rolled copper depends on its surface treatment and the resin system because of the lack of "teeth" present on electrodeposited copper. Rolled copper is commonly used in FWBs.

The use of rolled-annealed copper foil is preferred because of the greater ductility. This reduces the possibility of microcrack formation during bending and folding. The copper is adhesively bonded to flexible substrate. In microwave applications, rolled copper is advantageous because of its smooth surface. The smoothness allows control of the dissipation factor through close tolerance on dielectric spacing. In pattern generation, which involves etching, a uniform surface produces cleaner lines because preferential etching along grain boundaries is minimized [Oswald and Miranda (1977)].

Electroless copper is a copper foil chemically deposited through a chemical reduction process. The deposition process progresses without the aid of an externally applied current, which differentiates it from electrodeposited copper.

Table 46. Properties of adhesives for flex circuits (Du Pont 1990)

Property	Test method	Acrylic	Epoxy	Polyester
Peel strength (N/cm)	IPC-TM-650 Meth-2.4.9			
As received		14.0	12.3	5.3
After solder		12.3	10.5	N.A
After temperature cycle		12.3	10.5	5.3
Low temperature flexibility	IPC-TM-650 Meth-2.6.18	Pass	Pass	N.A
Flammability (Oxygen index)	IPC-TM-650 Meth-2.3.8	15	18	18
Solder float	IPC-TM-650 Meth-2.4.13A	N.A	PASS	N.A
	Meth-2.4.13B	N.A	N.A	N.A
Dielectric constant @ 1 MHz	IPC-TM-650 Meth-2.5.5.3	4	4	4.6
Loss tangent @ 1 MHz	IPC-TM-650 Meth-2.5.5.3	0.05	0.06	0.13
Volume resistivity (MΩ-cm)	IPC-TM-650 Meth-2.5.17	10^6	10^6	10^6
Surface resistivity (MΩ/sq.)	IPC-TM-650 Meth-2.5.17	10^6	10^3	10^4
Dielectric strength (V/mil)	ASTM-D-149	1000	500	1000
Insulation resistance (MΩ)	IPC-TM-650 Meth-2.6.3.2	10^4	10^2	10^3
Moisture absorption (wt. %)	IPC-TM-650 Meth-2.6.2	6	4	2
Fungus resistance	IPC-TM-650 Meth-2.6.2	Non-nutrient	Non-nutrient	Non-nutrient

Table 47. Metal foil properties

Metal foil	Tensile strength (MPa)	Elasticity Modulus (GPa)	Elongation (%)	Fatigue ductility (%)	CTE (ppm/°C)	Thermal conductivity (W/m-°C)
Copper (rolled, annealed)	235	117	20.0	65	16.6	392
Copper (electro-deposited)	310	110	12.0	35	16.6	392
Aluminum	83	70	18.0	-	23.6	155
Nickel (rolled)	495	205	40.0	-	13.3	125

Table 48. Copper foil classes

Class	Type	Designation	Description
1	E	STD	Standard electrodeposited
2	E	HD	High-ductility electrodeposited
3	E	THE	High-temperature elongation electrodeposited
4	E	ANN	Annealed electrodeposited
5	W	AR	As rolled-wrought
6	W	LCR	Light cold rolled-wrought
7	W	ANN	Annealed-wrought
8	W	ARLT	As rolled-wrought, low-temperature annealable

Table 49. Copper foil minimum tensile strength

Class	Weight (oz)	Tensile Strength (MPa)	
		At 23°C	At 180°C
Foil type E			
1	½	103.35	N/A
	1	206.70	N/A
	2+	206.70	N/A
2	½	103.35	N/A
	1	206.70	N/A
	2+	206.70	N/A
3	½	103.35	-
	1	206.70	137.8
	2+	206.70	172.2
4	1	137.80	103.3
	2+	137.80	103.3
Foil type W			
5	½	344.50	-
	1	344.50	137.80
	2+	344.50	375.80
6	1	172.25-344.5	N/A
	2+	According to temperature	N/A
7	½	103.35	-
	1	137.80	96.46
	2+	172.25	151.58
8	½	103.35	N/A
	1	137.80	N/A

Table 50. Application of copper foils

Maximum strain range (%)/Minimum bend diameter (mil) accommodated by 1 oz foil[a]

Type	Class	At room temperature (23°C) Flex to Install				At elevated temperature[b] (180°C) Flex to Install		
		Handling	Single bend	Low cycle fatigue	Continuous flexing, high cycle fatigue	Single bend	Low cycle fatigue	Continuous flexing, high cycle fatigue
E	1	Good	15/7.9	4.4/63	0.19/1450	N/A	N/A	N/A
E	2	Good	30/3.3	7.1/38	0.21/1340	N/A	N/A	N/A
E	3	Good	20/5.6	5.3/51	0.20/1410	15/7.9	4.2/65	0.14/2050
E	4	Caution	60/1.0	11.5/23	0.18/1550	30/3.3	6.9/39	0.12/2240
W	5	Good	30/3.3	7.5/36	0.32/870	15/7.9	42/65	0.14/2050
W	6	Refer to footnote d						
W	7	Caution	65.0/0.8	12.3/21	0.19/1510	45/1.7	9.3/29	0.13/2090
W	8	Good	25/4.0	6.2/44	0.15/1890	N/A	N/A	N/A

[a]Large strain range and smaller diameter values indicate superior performance for a given strain mode.

[b]The values for elevated temperature applications should be used for qualitative purpose.

[c]Low-cycle fatigue <500 cycles to failure; high-cycle fatigue >10^4 cycles to failure.

[d]Choice of temper. allows tradeoffs in handling/high-cycle fatigue versus/low-cycle fatigue properties.

4.6   Conformal Coatings

Conformal coatings are applied to circuit board assemblies to protect interconnect conductors, solder joints, the board itself, and components. The coatings provide a semi-hermetic barrier to reduce solubility and permeability of moisture, hostile chemical (corrosive) vapors, and solvents in the coating. Use of conformal coatings avoids chances of dendritic growth between conductors, conductor bridging from moisture condensation, and reduction in insulation resistance by water absorption. The properties of interest for conformal coatings are resistance to chemicals, moisture, and abrasion. Other properties with significance in particular applications are flexibility, modulus of elasticity, thermal conductivity, CTE, and the useable temperature range. The dielectric constant and loss tangent of the conformal coating becomes important in high-speed applications. The material must be selected to ensure that it lasts the life-time of the product, is easy to apply and rework, and is cost effective.

Polyimides, polyamide-imides, silicones, acrylics, polyurethanes, fluoropolymers, parylenes, and epoxies are the typical materials used for conformal coatings. Except for parylenes, which are vapor-deposited, the coatings are usually applied in liquid form and are cured using infrared (IR) or ultraviolet (UV) radiation. UV curable polymers include acrylics and epoxies. IR-curable polymers include silicones, polyurethanes, and butyl rubber.

Table 51. Comparison of coating materials (Lampack 1989)

Characteristics	Urethane	Acrylic	Silicone	Epoxy	Parylene
Application	1	1	3	2	4
Chemical removal	2	1	2	5	5
Burn removal	2	1	5	4	5
Mechanical Removal	2	2	1	5	3
Abrasion resistance	2	2	3	1	2
Adhesion to various Substrates	2	3	4	1	2
Humidity resistance	1	1	2	4	1
Long-term Humidity exposure resistance	1	2	3	4	1
Thermal shock Resistance	2	3	1	5	1
Mechanical Strength	2	3	4	1	2
Insulation Properties	1	1	2	3	1
Dielectric Properties	1	1	2	3	1

1 = most favorable; 5 = least favorable

A qualitative comparison of the different coating materials is given in Table 51. Some of the essential characteristics of conformal coating materials are given in Table 52.

Polyurethanes are available as one component and two component resins. The cured properties of one- and two-component systems are very similar. The difference is in their method of application. Polyurethanes offer good moisture, fungus, abrasion, solvent, and chemical resistance. They have good adhesion, low shrinkage, flexibility, and elasticity, and are particularly suited to applications requiring humidity resistance. Rework is difficult, due to their high chemical resistance. They are unsuitable for high-frequency and high-temperature applications, due to the strong dependence of their dielectric constant and loss tangent on these variables. Many polyurethane systems contain isocyanate groups, which can be very hazardous to health.

Acrylics (PMMA) have excellent moisture resistance, dielectric properties, and reworkability, but they have poor abrasion resistance. They have excellent arc resistance, even after long-term immersion in water or exposure to humidity. They are resistant to attack by weak acids and bases, but are attacked by solvents and strong acids. Reworkability is due to their reduced solvent resistance. Decreasing dielectric constants with increasing frequencies make acrylics attractive candidates for high-frequency applications.

Epoxies provide excellent chemical, moisture, and abrasion resistance. The flexibility required for the coating is obtained by controlling the degree of cross-link density among the molecules through reduced amounts of the curing agent. Stress-sensitive components need to be coated with a compliant layer before applying an epoxy coating to prevent fracturing due to shrinkage of the epoxy upon polymerization. Rework on an epoxy-coated PWB requires the application of heat or abrasion for removal of the coating because solvents will also dissolve the epoxy if they are present in the PWB laminate.

Silicone-based coatings are well suited for high-temperature and high-speed applications. They are flexible, tough, and have high-temperature resistance and low dielectric constants. They resist thermal and oxidative deterioration, have good surface resistance, and are fungus- and flame-resistant. Their low surface tension helps in water and moisture resistance. However, they possess a high CTE and poor adhesion. A primer coating can be applied to enhance adhesion. Silicones are available in solvent dissolvable form.

Polyimide coatings provide excellent high-temperature and chemical resistance. However, polyimide coatings require a high-temperature cure and are difficult to rework because of their high thermal stability and chemical resistance.

Table 52. Typical characteristics of coating materials

Property	Urethane	Acrylic	Epoxy	Silicone	Polyimide	DAP
Volume resistivity (Ω-cm) at 50% RH and 23°C	$10^{11} - 10^{14}$	10^{15}	$10^{12} - 10^{17}$	$10^{10}-10^{12}$	10^{16}	2×10^{16}
Dielectric constant at 1 MHz	4.2–5.1	2.2–3.2	3.3–4.0	2.6–2.7	3.4	3.4
Dissipation factor at 1 MHz	0.05–0.07	2.5–3.5	0.030–0.050	0.001–0.002	0.005	0.011
Thermal conductivity (W/m-°C)	0.07–0.31	0.12–0.25	0.17–0.21	0.15–0.31	-	0.17–0.21
CTE (ppm/°C)	100–200	50–90	45–65	60–90	40–50	-
Resistance to maximum continuous temperature (°C)	121	121	121	204	260	177
Chemical resistance to:						
weak acids	Slight to dissolve	None	None	Little or none	Resistant	None
weak alkalis	Slight to dissolve	None	None	Little or none	Slow attack	None
organic solvents	Resists most	Attacked by ketones, aromatics, and chlorinated hydrocarbons	Generally resistant	Attacked by some	Very resistant	Resistant

Table 53. Typical properties of Parylene coatings (Beach and Olson 1989)

Properties	Parylene-N	Parylene-C	Parylene-D
Density (g/cm^3)	1.11	1.289	1.418
Tensile modulus (GPa)	2.4	3.2	2.8
Tensile strength (MPa)	45	70	75
Yield strength (MPa)	42	55	60
Elongation to break (%)	30	200	10
Yield elongation (%)	2.5	2.9	3
Hardness (Rockwell R scale)	85	80	-
Melting point (°C)	420	290	380
CTE at 25°C (ppm/°C)	69	35	-
Specific heat capacity at 25°C (J/g-°C)	1.3	1.0	-
Thermal conductivity at 25°C (W/m-°C)	0.12	0.082	-
Dielectric constant			
at 60Hz	2.65	3.15	2.84
at 1 kHz	2.65	3.10	2.82
at 1 MHz	2.65	2.95	2.80
Loss tangent			
at 60 Hz	0.0002	0.02	0.004
at 1 kHz	0.0002	0.019	0.003
at 1 MHz	0.0006	0.013	0.002
Dielectric strength, 25 μm thick (MV/m)			
sort time	275	220	215
step-by-step	235	186	-
Volume resistivity at 23°C and 50% RH (MΩ)	14	8.8	2
Surface resistivity at 23°C and 50% RH (Ω)	10^{13}	10^{14}	10^{16}
Water absorption (%)	<0.1	<0.1	<0.1
Gas permeability at 25°C (mol/ Pa-s-m)			
N_2	15.4	2.0	9.0
O_2	78.4	14.4	64.0
CO_2	429.0	15.4	26.0
H_2S	1590	26.0	2.9
SO_2	3790	22.0	9.53
Cl_2	148	0.7	1.1

Diallyl phthalate(DAP) coatings have the advantage of low shrinkage upon curing. The disadvantages include high-temperature cure and low resistance to acids and alkalis. Rework is difficult due to DAP's resistance to high temperatures and most organic solvents.

Parylenes have excellent moisture and chemical resistance and good electrical and mechanical properties. Parylenes are noted for their ability to penetrate small spaces due to vapor deposition. Uniform thickness over edges, corners, and depressions is an added advantage of the deposition process. Further, the coating is nearly pinhole-free, even for coatings as thin as 0.004 mil. Removal of the coating requires heat softening and abrasion with a brush or microabrasive blasting, and is therefore complex. Commercially, parylene is available as parylene-N, parylene-C, and parylene-D. Parylene-N is the most stable in electrical properties with frequency and temperature. Table 53 lists the properties of some parylene conformal coating resins.

5   THIRD-LEVEL PACKAGING MATERIALS

The third level of packaging includes the interconnections and hardware required to realize an electronic system after the printed wiring boards (PWBs) have been assembled. The interconnections include connections between daughter boards, between subassemblies, between systems, and to peripherals. The hardware includes enclosures housing the PWBs and the cooling system. Required electrical interconnections are primarily achieved through the use of backpanels, connectors and cables. In the following sections the materials commonly employed for backpanels, connectors, and cables are discussed, with emphasis on the properties of importance to electronic applications.

5.1   Backpanel Materials

A backpanel is a board that interconnects printed wiring board assemblies and also serves as an input/output panel. When the interconnection density is low, it is cost-effective to employ a different version of backpanel known as a wire wrap board, or to use flexible cables instead of a multi-layer backpanel. In the construction and fabrication of a backpanel, multi-layer PWBs and their manufacturing technology are employed to realize high signal interconnection density and power distribution requirements.

Materials used to fabricate a typical backpanel are the same as those employed for multi-layer PWBs. These materials are often fiber-reinforced composites (see Section 4). For low-connection-density and high-power applications, metal plates are employed as wire wrap boards for point-to-point signal connection. The metal plates are used as ground and power planes. Good thermal conductivity and high current-carrying capability are beneficial for metals in low-voltage and high-current applications. A condenser-type construction of the panel smooths small fluctuations in power. High dielectric constant materials are used between the plates to enhance the capacitance between the ground plate and the power plate.

5.2   Connector Materials

Connectors are available in various designs and architectures, depending on the performance, reliability, and geometric configuration requirements. The materials employed in the fabrication of a typical connector include insert materials and pin-and-contact metals. Plating and/or lubricants are often used to ease assembly, reduce wear, and to improve the contact resistance between pin and contact.

Insert materials: The insert material provides insulation and support to the pins and contacts. To ensure proper mating and contact force, especially in the case of high-pin-density connectors, pin-to-pin dimensional stability despite

Table 54. Ranking of thermosetting polymers by key property values (Shugg 1986)

Thermosetting polymer	A	B	C	D	E	F	G	H	I	J	K	L
Alkyd/polyester	5	4	3	3	1	2	1	5	3	5	2	1
Allyl	1	1	1	1	1	5	3	1	1	2	4	5
Melamine	3	5	5	3	5	3	5	4	3	3	3	1
Epoxy (Novolac)	3	2	2	1	1	1	3	3	3	1	5	1
Phenolic	2	3	4	3	1	4	2	2	2	4	1	1

1 = most favorable; 5 = least favorable
A: specific gravity; B: dielectric strength; C: dielectric constant; D: dissipation factor; E: volume resistivity; F: arc resistance; G: water absorption; H: deflection temperature; I: maximum service temperature; J: tensile strength; K: izod impact strength; L: flammability (1 = V–0 and 5 = HB)

temperature fluctuations is important. Therefore, materials with low CTEs and low mold shrinkage are preferred. The reduction in pin-to-pin distance with increasing pin density requires high volumetric and surface resistivity for proper electrical performance. High thermal conductivity, low density, and resistance to corrosion, water absorption, chemical attack, and flame are desirable properties for an insert material, although knowledge of the mission profile would be required to determine their relative importance. Flexibility is essential for a snap-fit connector design.

In high-speed applications with subnanosecond rise times and clock rates in the megahertz range and beyond, the impedance of the connector must be matched with the incoming and outgoing signal paths.

In such applications, the signal conductors in the connector act as transmission lines and the dielectric constant of the insert material actively determines performance. A low dielectric constant material between the signal and ground plane/pin controls cross-talk and achieves high-speed signal transmission. However, high-dielectric-constant materials are desirable between the ground and power plane to achieve good capacitance and smooth small perturbations in power.

Polymers are common materials used as insert materials for pins and contacts. Figure 4 presents their classification as thermosets and thermoplastics. Thermosets include alkyds, allylics (DAP), aminos, epoxies, and phenolics. Thermoplastics are subdivided into amorphous and crystalline; both kinds are employed in electronic packages. Among the amorphous thermoplastics are ABS, modified phenylene-oxide-based resin, polycarbonate, and polystyrene. The crystalline thermoplastics include acetals, nylon, polypropylenes and polyesters. Generally, thermosets are not recommended for high operating temperatures since they tend to embrittle. For low-temperature applications, thermosets can be easily modified to retain their flexibility when thermoplastics are not recommended due to their poor flexibility. A comparative ranking of thermosets and thermoplastics by properties is provided in Tables 54 and 55. Key properties of the insert materials are covered in Table 56.

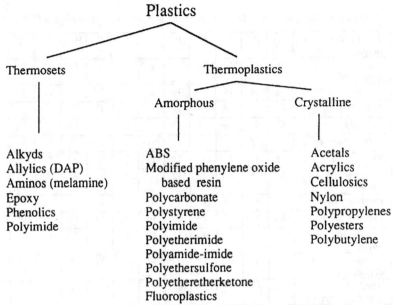

Figure 4: Classification of plastics as thermosets and thermoplastics.

Polysulphones are high-temperature thermoplastics. In general, polysulphones are transparent and exhibit high rigidity, low creep, excellent thermal stability, and good flame resistance compared with other thermoplastics. The glass transition temperature is often in excess of 200°C. The loss tangent is very low, and they also possess a low dielectric constant that is stable over a broad range of temperatures and electromagnetic frequencies. The major weakness of polysulphones is their low chemical resistance. Although resistant to most solvents, acids, and alkalis, they exhibit stress cracking in the presence of organic ketones, esters, and chlorinated hydrocarbons. Examples of polysulphones include Radel and Udel from Union Carbide and Victrex from ICI.

Polyphenylene sulphides (PPS) are thermoplastics with resistance to almost all organic solvents, even at elevated temperatures. Polyphenylene sulphides are flame resistant and have excellent electrical insulation properties. Their melt viscosity can be easily varied, which is advantageous for injection molding of connectors.

Phenolic resins are low-cost thermosets and are widely available in different grades. Pure phenolic resin is very brittle, but when used in a fiber composite, the composite exhibits improved impact resistance. Phenolics are poor thermal conductors, do not support combustion, are hard and strong, do not absorb water, and are resistant to mild acids and common solvents. Military specification MIL-M-14G grade MFG and GPI-100 possess good electrical properties.

Table 55. Ranking of thermoplastic polymers by key property values (Shugg 1986)

Thermoplastic polymer	A	B	C	D	E	F	G	H	I	J	K
Acetal homopolymer	18	15	14	7	15	4	11	8	14	6	8
Acetal copolymer	17	15	14	8	16	2	9	9	9	12	7
Nylon 6/6	8	1	18	18	18	7	18	15	7	4	6
Nylon 11	4	8	16	16	16	9	13	17	17	13	13
Polycarbonate	10	15	9	14	4	11	7	6	8	11	1
Polysulfone	11	8	7	5	6	10	13	5	5	5	8
Polyphenylene sulfide	14	15	8	2	12	18	2	2	2	9	18
Polypropylene	1	1	1	4	9	6	1	16	9	18	14
Polystyrene	4	5	2	1	9	16	2	11	16	16	15
Polystyrene-butadiene	2	12	2	9	12	17	4	11	17	16	5
Styrene-acrylonitrile	7	8	4	13	9	12	8	11	15	9	15
Acrylonitrile-butadiene-styrene	2	8	4	12	8	13	9	10	12	15	2

1= most favorable; 15= least favorable
A: specific gravity; B: dielectric strength; C: dielectric constant; D: dissipation factor; E: volume resistivity; F: arc resistance; G: water absorption; H: deflection temperature; I: maximum service temperature; J: tensile strength; K: izod impact strength

Type MFG is a mineral-filled compound with low loss tangent, high dielectric strength, and low water absorption. Improved impact strength, along with good electrical properties, are obtained by glass-fiber filling in grade GPI-100.

Amino resins used for electronic packaging focus on melamine, because most amino compounds suffer from shrinkage during and after molding. The shrinkage leads to cracks around pins with the passage of time. Aminos also have high water absorption and thermal expansion. Melamine possesses better dimensional stability and resistance to water absorption and chemical attack than other amino compounds. Melamine is a good electrical insulator, resistant to scratch, arc, and solvents. Melamine is decomposed by strong acids and attacked by alkalis.

Epoxy resins are thermoset polymers with dimensional stability, low dielectric constant, low loss tangent, and high tensile strength. The flexibility of epoxies is controlled by the extent of cross linking. Cured epoxy properties are maintained under high moisture and temperature conditions. Epoxy resins are chemically resistant and are affected only by strong acids and ketones. Their low viscosity, low shrinkage, and thermal stability permit the manufacture of high-density connectors.

Diallyl phthalate (DAP) is frequently used in the connector industry because of its high arc resistance, low dielectric constant, high dielectric strength, and good thermal and mechanical properties. DAPs exhibit dimensional stability, with shrinkage less than 3%, which makes them ideal for close-tolerance applications. DAPs maintain high insulation resistance under both thermal stress and high humidity. Glass-fiber reinforced DAPs are military-approved molding materials for use in connectors.

Nylon is a generic name for the family of polyamides. Nylons exhibit high impact toughness, and tensile and flexural strengths over a wide temperature range. As electrical insulators, they have good dielectric strength and high resistivity. However, they have a high dielectric constant, which limits their use to lower speed applications. Nylon resins also absorb moisture up to 2.5% by weight. The absorbed moisture induces swelling and increases the dielectric constant. Simultaneously, the absorbed moisture improves the tensile strength and the impact toughness.

Acetals are thermoplastics with mechanical properties similar to metals such as aluminum, brass, and zinc. They have good dimensional stability, a high tensile strength, flexural strength, rigidity, resilience, and toughness not obtainable in most other thermoplastics. Electrically, they have a low dielectric constant and loss tangent over a wide range of both frequencies and elevated temperatures. Acetals exhibit high moisture absorption, but their electrical properties do not significantly degrade.

Polycarbonate (PC) resins have good flexibility, low dielectric constant, high volumetric resistance, exceptional impact toughness, and excellent dimensional stability over a wide range of temperature and humidity. PCs have outstanding thermal stability, coupled with predictable shrinkage. The resins are self-extinguishing. However, the permeability of polycarbonate to moisture and gases is relatively high.

ABS is a polymer in which copolymer acrylonitrile-styrene is grafted on butadiene polymer. A wide range of properties is obtainable by varying proportions of monomers acrylonitrile, butadiene and styrene. Acrylonitrile monomer imparts heat stability and chemical resistance; butadiene contributes to low-temperature property retention and toughness; and styrene is added for processing ease. Flame retardancy is achieved by halogenation or by alloying with poly-vinyl-chloride. ABS resins have excellent impact toughness at temperatures as low as − 40°C. Their low specific gravity combined with high toughness, rigidity, and good flow properties are useful in electronic applications. ABS has poor resistance to ultraviolet light.

Neoprene provides environmentally sealed connectors when used in ambient temperatures from –55 to 85°C. Neoprene is oil-resistant, molds well, and is among the most heat resistant of the organic rubbers. Silicone rubbers are generally used when service conditions require temperatures above 175°C or in cold weather. Fluorosilicone rubber is used where maximum fuel and oil resistance is required and temperatures range between –20 and 260°C.

Table 56. Key properties of encapsulants (Shugg 1986)

Property	ASTM test method	A	B	C	D	E	F	G
Specific gravity @ 23° C	D 792	1.92	1.74	1.92	1.70	1.14	1.04	1.20
Dielectric strength at 23°C, 0.125", Short time (kV/mm)	D 149	11.8	15.0	15.3	15.7	23.6 (dry)	29.5 (dry)	15.0
Dielectric constant, 23°C, 1 MHz	D 150	6.2	5.0	4.6	3.5	4.6	3.1	3.0
Dissipation factor@ 23°C 1 MHz	D 150	0.02	0.02	0.01	0.01	0.10	0.05	0.01
Volume resistivity, 23°C (Ω-cm)	D 257	2e11	2e11	1e13	1e13	1e13	1e14	8.2E16
Arc resistance, 23°C (s)	D 495	180+	180	187	140	130	123	120
Water absorption, 1/8" thk, 24 hr, 23 °C (wt %)	D 570	0.40	0.15	<0.2	<0.2	1.2	0.3	0.15
Deflection temperature, @ 264 psi (°C)	D 638	227	260+	232	282	90	55	129
Maximum service temperature (°C)		204	232	204	260	130	65	115
Tensile strength, 23°C (MPa)	D 638	55.1	51.7	72.4	68.9	77.2	56.5	62.0
Izod impact strength, 1/8" thk. (m-N/m)	D 256	294	641	24	267	112	40	641
Flammability:								
UL Standard 94 rating,1/16" specimen	D 2863							
Standard grades		V-0	V-0	V-0	HB	V-2	V-2	V-2
FR grades		-	-	-	V-0	V-0	None	V-0
Oxygen index (%)								
Standard grades		N/A	45-60	-	26-32	-	-	-
FR grades		-	-	-	36	-	-	-

A: Glass-filled melamine MIL-M-14G Type MMI-30; B: Glass-filled phenolic MIL-M-14G Type GPI-100; C: Glass-filled epoxy; D: Glass-filled alloy MIL-M-14G Type GPI-30; E: Nylon 6/6; F: Nylon 11; G: Polycarbonate.

Table 57. Copper base metal alloys (Ginsberg 1979)

Property	Alloy 260[a] Brass	Alloy 172[a] Beryllium-copper[c]	Alloy 510 Grade A Phosphor-bronze	Alloy 638	Alloy 725	Alloy 762 Nickel-silver
Nominal composition	Cu 70 Zn 30	Cu 98.1 Be 1.9	Cu 94.81 Sn 5.0	Cu 95 Al 2.8 Si 1.8 Co 0.4	Cu 88.2 Ni 9.5 Sn 2.3	Cu 59.25 Zn 28.75 Ni 12
Electrical conductivity at 20 °C (Mmho/cm)	0.163	0.128	0.087	0.058	0.064	
Thermal conductivity at 20 °C (W/m-°C)	121	109-130	68.6	40.6	54.4	41.8
Density at 20 °C (g/cm^3)	8.54	8.260	8.86	8.29	8.89	8.70
Modulus of elasticity (GPa)	112	130	112	117	135	127
Yield strength, 0.2 % offset (MPa)						
Annealed	70 – 220	109	150	410 – 470	180	200
Half hard	290 – 410	123	330 – 480	530 – 630	400 – 510	410 – 580
Hard	460 – 530	127	520 – 620	640 – 720	520 – 560	580 – 680
Spring	580 – 630	N/A	650 – 760	700 – 790	550 – 650	710 – 770
Extra spring	600 – 690	N/A	690 – 770	750 min.	630 – 720	720 min.

[a]All property data for beryllium-copper is for material after age hardening heat treatment.
[b]Designation of Copper Development Association.

Table 58. Properties of nickel-based alloys (Ginsberg 1979)

Alloy	Nominal Composition (%)	Tensile strength (MPa)	Yield strength (MPa)	Elongation (%)	Modulus of elasticity (GPa)	Electrical conductivity (MΩ -cm)
Nickel 200	99 Ni	379–758	103–689	55–10	206	0.106
Nickel 270	99.97 Ni	344–655	103–620	50–4	206	0.134
Duranickel 301	93 Ni 4.5 Al	620–1448	209–1206	55–15	206	0.024
Be-Ni 440	98 Ni 2 Be	655–1861	275–1586	30–8	186–207	0.027

Pin and contact materials: Properties required of metals used in fabricating the conductor pins and contacts are low electrical resistivity, high strength, a high modulus of elasticity, creep resistance, wear resistance, oxidation resistance, and corrosion resistance. Applications in which contact force is due to spring action require high yield strength and high tensile modulus with minimal creep and stress relaxation. Ease of fabrication by stamping and bending sheets requires formability and ductility. A low dielectric constant is required for high-speed connectors to achieve high-density interconnection with controlled impedance and crosstalk. Generally, copper- or nickel-based alloys are used for pins and contacts.

When copper is alloyed with zinc, it is usually called brass. If it is alloyed with another element, it is often called bronze. Sometimes the element is also specified as, for example, in phosphor bronze. Brass, beryllium-copper, phosphor-bronze, and copper-nickel are commonly employed for both pins and contacts to provide spring characteristics along with high electrical conductivity and other wear characteristics. Table 57 lists properties of some copper alloys used for pins and contacts.

Beryllium-copper is a common contact material due to its good spring properties and corrosion resistance. Beryllium-copper exhibits high strength, hardness, fatigue endurance and wear resistance. The yield strength is usually in the range of 550 to 1137 MPa. Due to the high strength of beryllium-copper, the spring contacts can be made small. Beryllium-copper (Be-Cu) can be formed in any direction without fracturing. Thus, complex shapes and sharper bends in pins and contacts can be easily formed. Be-Cu also has high resistance to anelastic (defined as time-dependent elastic strain) creep behavior and resistance to stress relaxation. Corrosion resistance, high electrical conductivity, the highest hardness among copper-base alloys and good wear resistance all contribute to make BeCu an ideal spring contact material for connectors.

Phosphor-bronze, made with a large percent of tin compared with the amount of phosphorous, is especially resistant to fatigue and corrosion. It has high tensile strength, high capacity to absorb energy, and wear resistance. Phosphor-bronze has good thermal and electrical conductivity, though not as high as beryllium-copper. The most popular phosphor-bronze composition is 95% copper, 4.75% tin, and 0.25% phosphorous.

Nickel-based alloys have advantages over copper-based alloys: formability before age hardening, improved stress-corrosion resistance, galvanic compatibility, resilience, low stress relaxation and the ability to clad or plate. Typical properties of some nickel alloys are listed in Table 58. The addition of nickel to brass makes nickel-silver, an alloy more resistant to stress, corrosion, and cracking than brass. Nickel-containing alloys generally have a high modulus of elasticity and fatigue strength, making them the preferred materials for applications requiring spring contact. The spring properties are retained at high operating temperatures due to low stress relaxation. At temperatures above 93°C, it is better to employ nickel alloys than copper alloys.

Beryllium-nickel is a nickel-base alloy containing 1.95 percent beryllium and 0.5 percent titanium. It is a soft material that hardens upon heat treatment due to precipitation hardening. It has excellent resistance to stress relaxation at elevated temperatures, a high modulus of elasticity, high fatigue strength, and stress corrosion resistance. The disadvantage of beryllium-nickel is that it has lower electrical conductivity than copper base alloys.

Table 59. Plating characteristics for contacts

Plating material	Plating characteristics
Silver	General-purpose plating for power contacts. Shelf life is poor. Will tarnish when exposed to atmosphere and contact resistance will increase. Problem with low-level circuits.
Gold	Excellent conductor, very stable material. Hard gold platings good for repeated insertions, has low contact resistance and excellent corrosion resistance.
Nickel	Excellent corrosion resistance, fair conductivity. Generally used as an undercoat where high temperatures would cause migration of silver through gold. Good wear resistance.
Rhodium	Used where exceptional wearing characteristics are required. Lower conductivity than gold or silver, very high cost.
Tin	Forms cold weld joint under weld, good conductivity, excellent solderability, low cost, poor wire resistance.
Gold over silver	Good for dry circuit applications, low contact resistance, moderate corrosion resistance. Not usually used.
Gold over nickel	Good for repeated insertions and excellent corrosion resistance.
Rhodium over nickel	Maximum wear resistance, high-temperature operation. Has higher contact resistance than other platings.

Pin and contact plating. Enhancing the mechanical connection between a pin and contact is often accomplished by plating relatively soft metal on the surfaces. Plating can also retard corrosion, preserve solderability, and guarantee operation of dry circuit contacts that are closed infrequently. Gold, tin, palladium, platinum, rhodium, and some palladium-nickel alloys are commonly employed for plating. Table 59 shows the plating characteristics of some of the most commonly used plating materials. Gold is an excellent conductor and is very stable. Its wear resistance and inertness to chemical attack and surface contamination make it a good plating material. Gold contact plating is typically in the range of 0.4 to 1.3 microns in thickness. Gold plating with hardness values of 140 – 180 Knoop has good wearability and ductility.

Tin is used as a low-cost plating material. Tin oxidizes easily and forms a film that is thin, hard, and brittle. However, with repeated engagement and disengagement of connectors, the tin oxide layer forms and breaks repeatedly, causing an increase in contact resistance. The usefulness of tin as a contact material lies in the fact that it is soft and has good conductivity and excellent solderability. The disadvantages include hard and brittle tin oxides that increase wear.

5.3   Cables and Flex Circuit Materials

A cable or a flex circuit is often employed to interconnect electronic circuits that are not easily connected by other means. The interconnection density obtained with cables and flex circuits is lower than that obtained using a backpanel. Properties desired in cable insulation and flexible circuit substrate materials include mechanical flexibility, fatigue endurance, and resistance to chemicals, water absorption, and abrasion. Because the total length of the cable or flex circuit could be very large, an insulator material with low density is required to reduce the total weight.

Cable conductors. Copper is by far the most widely used conductor material. It has high electrical conductivity, thermal conductivity, solderability, and resistance to corrosion, wear, and fatigue. Annealed copper conductors can best withstand the flex and vibration stresses normally encountered in use. Copper-covered steel combines the corrosion resistance of copper with the strength of steel, and is used for high-frequency applications.

Cable insulators. Both thermoplastics and thermosets are employed as cable insulator materials. Mechanical and electrical properties of insulating materials are listed in Tables 60 and 61, respectively. Thermoplastic materials possess excellent electrical characteristics and are available at relatively low cost. Compared to thermosets, thermoplastics can be used to obtain good electrical properties with less material. Some of the important cable materials include polyvinyl   chloride   (PVC)   compounds,   polyamides,   polyethelenes, polypropylenes, polyurethanes, and fluoropolymers.

Table 60. Mechanical properties of cable insulating materials (Harper 1991)

Insulation	Common designation	Tensile strength (MPa)	Elongation (%)	Specific gravity	Abrasion resistance	Cut-through resistance	Temperature resistance (mechanical)
Polyvinyl chloride	PVC	16.6	260	1.2–1.5	poor	poor	fair
Polyethylene	PE	9.6	300	0.92	poor	poor	good
Polypropylene	PPE	41.4	25	1.4	good	good	poor
Cross-linked polyethylene	IMP	20.7	120	1.2	fair	fair	good
Polytetrafluoroethylene	TFE	20.7	150	2.15	fair	fair	excellent
Fluorinated ethylene propylene	FEP	20.7	150	2.15	poor	poor	excellent
Monochlorotrifluoroethylene	Kel-F	34.5	120	2.13	good	good	good
Polyvinylidene fluoride	Kynar	49.0	300	1.76	good	good	fair
Silicone rubber	Silicone	5.5–12.6	100–800	1.15–1.38	fair	poor	good
Polychloroprene rubber	Neoprene	1.0–27.6	60–700	1.23	good	good	fair
Butyl rubber	Butyl	4.8–10.3	500–700	0.92	fair	fair	fair-good
Fluorocarbon rubber	Viton	16.6	350	1.4–1.95	fair	fair	fair-good
Polyurethane	Urethane	34.5–55.2	100–600	1.24–1.26	good	good	fair-good
Polyamide	Nylon	27.6–48.3	300–600	1.10	good	good	poor
Polyimide film	Kapton	124.1	707	1.42	excellent	excellent	good
Polyester film	Mylar	89.6	185	1.39	excellent	excellent	good
Polyalkene	-	13.8–48.3	200–300	1.76	good	good	fair-good
Polysulfone	-	69.0	50–100	1.24	good	good	good
Polyimide-coated TFE	TFE/ML	20.7	150	2.2	good	good	good
Polyimide-coated FEP	FEP/ML	20.7	150	2.2	good	good	good

Table 61. Electrical properties of cable insulating materials (Harper 1991)

Insulation	Common designation	Dielectric strength (kV/mm)	Dielectric constant @ 1 kHz	Dissipation factor @ 1 kHz	Volume resistivity (Ω-cm)
Polyvinyl chloride	PVC	15.8	5-7	0.02	2×10^{14}
Polyethylene	PE	18.9	2.3	0.005	10^{16}
Polypropylene	-	29.5	2.54	0.006	10^{16}
Cross-linked polyethylene	IMP	27.6	2.3	0.005	10^{16}
Polytetrafluoroethylene	TFE	18.9	2.1	0.0003	10^{18}
Fluorinated ethylene propylene	FEP	19.9	2.1	0.0003	10^{18}
Monochlorotrifluoroethylene	Kel-F	17.0	2.45	0.025	2.5×10^{16}
Polyvinylidene fluoride	Kynar	50.4(8mil)	7.7	0.02	2.0×10^{14}
Silicone rubber	Silicone	22.6–27.6	3.0-3.6	0.003	2.0×10^{15}
Polychloroprene rubber	Neoprene	32.0	9.0	0.030	10^{11}
Butyl rubber	Butyl	23.6	2.3	0.003	10^{17}
Fluorocarbon rubber	Viton	19.7	4.2	0.14	2.0×10^{13}
Polyurethane	Urethane	17.7–19.7	6.7-7.5	0.055	2.0×10^{11}
Polyamide	Nylon	15.2	4-10	0.02	4.5×10^{13}
Polyimide film	Kapton	212.6(8mil)	3.5	0.003	10^{18}
Polyester film	Mylar	102.4	3.1	0.15	6.0×10^{16}
Polyalkene	-	73.6	3.5	0.028	6.0×10^{13}
Polysulfone	-	16.73	3.13	0.0011	5.0×10^{16}
Polyamide-coated TFE	TFE/ML	18.9	2.2	0.0003	10^{18}
Polyimide-coated FEP	FEP/ML	18.9	2.2	0.0003	10^{18}

PVC compounds are widely used in cables because of their high dielectric and mechanical strength, flexibility, and flame, water, and abrasion resistance. PVC compounds are easy to process; however, their processing is a health hazard.

Nylon suffers from poor moisture absorption, which degrades electrical properties. Consequently it is not used as a primary insulator. However, nylon has good mechanical, thermal, and chemical properties, which make it a good jacket material over other insulating materials. Polyethylene and polypropylene are used for high-speed applications, where a low dielectric constant and low loss tangent are necessary in the cable jacket material. At low temperature, these materials are stiff, although bendable without breaking. They are also resistant to water absorption, chemical attack, heat, and abrasion. Good electrical properties help reduce the insulation thickness, which is required to attain high-density connections.

Polyurethanes possess outstanding abrasion resistance and exhibit resistance to certain chemicals that attack other insulation materials. They are used where abrasion resistance is of prime concern.

SUMMARY

The ever-increasing demands placed on electronic devices with respect to performance, reliability, manufacturability, and cost challenge the capabilities of packaging materials in terms of their properties. New materials with advanced and often tailored properties are finding multiple applications at various levels of electronic packaging. However, with the use of these materials comes the added responsibility of assessing the manufacturing and reliability issues associated with them and their interfaces. Success in the area of electronic packaging begins with the proper choice of packaging material, based on an understanding of the material's behavior and the influence of environmental and operating conditions on the material's properties. Experimental characterization, numerical simulation, and analytical modeling of materials are all fundamental to the proper utilization of electronic packaging materials.

APPENDIX A:

NOMENCLATURE

ε	complex electrical permittivity (= , N - , O) (dimensionless)
ε'	dielectric constant (dimensionless)
ε''	loss factor (dimensionless)
$\tan \delta$	dissipation factor or loss tangent (dimensionless)
V_{DS}	dielectric strength (MV/cm)
ρ_v	volume resistivity (MS-cm)
ρ_s	surface resistivity (S per square)
IR	insulation resistance (nA)
Δt_{arc}	arc resistance (s)
k	thermal conductivity (W/m-EC)
T_d	deflection temperature (EC)
T_g	glass transition temperature (EC)
CTE	coefficient of thermal expansion (ppm/EC)
σ	stress (MPa)
$\in$	strain (dimensionless)
E	modulus of elasticity (GPa)
σ_{ut}	tensile strength (MPa)
σ_{uc}	compressive strength (Mpa)
σ_u	ultimate strength (Mpa)
v	Poisson's ratio (dimensionless)
FM	flexural modulus (Gpa)
K_{IC}	fracture toughness under plane strain conditions (Mpa%m)
σ_{cs}	creep strength (MPa)
σ_{cu}	creep rupture strength (Mpa)
S_N	fatigue strength at N cycles (MPa)
S_e	endurance limit (MPa)
Δm	water absorption (weight %)
OI	oxygen index (%)
SG	specific gravity (dimensionless)
ρ	density (g/cm^3)
m_t	toxicity (Dg/m^3)

APPENDIX B:

LIST OF FIGURES:

REFERENCES

Agarwal, R.K., Dasgupta, A., Pecht, M., and Barker, D. (1991a). "Prediction of PWB/PCB Thermal Conductivity," *International Journal for Hybrid Microelectronics*, 14(3), 83-95.

Agarwal, R.K., Dasgupta, A., and Pecht, M. "Orthotropic Dielectric Constant and Loss Tangent of Plain-Weave Fabric Composites," submitted to *ASME Journal of Electronic Packaging*, January(1991b).

ANSI/IPC-EG-140. "Specification for Finished Fabric Woven from "E" Glass for Printed Boards." IPC, 7380 N Lincoln Ave., Lincolnwood, IL 60646(1988).

Atsumi, K., Ando, T., Kobayashi, M., and Usuda, O. (1986). "Ball Bonding Technique for Copper Wire," *36th Proceedings of IEEE Electronic Components Conference*, Seattle, Washington, May 5-7, 312-317.

Baker, J.D., Nation, B.J., Achari, A., and Waite, G.C. (1981). "On the Adhesion of Palladium Silver Conductors Under Heavy Aluminum Wire Bonds," *International Journal for Hybrid Microelectronics*, 4, 155-160.

Beach, F.B. and Olson, R. (1989). "Parylene Coatings, Electronic Materials Handbook," v. 1, Ed. M.L. Minges, ASM International, Materials Park, OH, 789-801.

Benoit, J. T., Grzybowski, R. R., and Kerwin, D.B. (1996). "Evaluation of Aluminum Wire Bonds for High Temperature (200°C) Electronic Packaging," *Proceedings of the Third International High Temperature Electronics Conference,* Albuquerque, NM, June 10-14, 1996. pp. III-17 to III-24.

Bhandarkar, S.M., Dasgupta, A., Barker, D., and Pecht, M. (1991). "Influence of Selected Design Variables on Thermo-Mechanical Stress Distributions in Plated-Through-Hole Structures," *ASME Journal of Electronic Packaging*, December.

Bhandarkar, S.M. (1991). Ph.D. Thesis, University of Maryland, College Park, Maryland.

Blood, B. and Casey, A. (1991). "Evaluating Multichip Module Packaging Technology," *PC Technology*, July.

Broek, D. (1986). *Elementary Engineering Fracture Mechanics*, Martinus Nijhoff, Boston, MA.

Clayton, A.M. (1989). "Epoxy Materials," Electronic Materials Handbook, v. 1, Ed. M.L. Minges, ASM International, Materials Park, OH, 825-837.

Corning. (1991). "Specialty Glass and Glass Ceramic Materials." Materials Business, Corning Inc., Corning, NY.

Dasgupta, A., Bhandarkar, S.M., Pecht, M., and Barker, D. (1990). "Thermoelastic Properties of Woven-Fabric Composites Using Homogenization Techniques," *Proceedings of the 5th Technical Conference of the American Society for Composites*, 1001-1010.

Dasgupta, A., Agarwal, R.K., and Pecht, M. (1991). "Prediction of Orthotropic Thermal Conductivity of Plain-Weave Fabric Composites Using Homogenization Technique," *Journal of Composite Materials*, November.

Dieter, G.E. (1986). *Mechanical Metallurgy*, 2nd ed., McGraw-Hill Book Company, NY.

Dosworth, R.S. (1991). New Development in TAB, *Surface Mount Technology*, March.

Du Pont (1986). *Data Manual for Kevlar 49 Aramid*, E.I. Du Pont de Nemours & Co., Inc., May.

Du Pont (1988). *Properties of Du Pont Industrial Filament Yarns*, Multifiber Bulletin X-272, July.

Du Pont (1990). *PYRALUX Flexible Composites*, E.I. Du Pont de Nemours & Co., Inc., January.

Fister, J., Breedis, J., and Winter, J. (1982). "Gold Leadwire Bonding of Unplated C194," *20th Proceedings of the IEEE Electronic Components Conference*, San Diego, California, 249-253.

Ginsberg, G.L., Ed. (1978). *Connectors and Interconnection Handbook*, 1-3, The Electronic Connector Study Group, Inc., Camden, NJ.

Gray, F.L. (1989). Thermal Expansion Properties, *Electronic Materials Handbook*, v. 1, Ed. M.L. Minges, ASM International, Materials Park, OH, 611-629.

Guiles, C.L. (1990). CTE Materials for PWB's, "*IPC Technical Paper 914,*" IPC Seminar Meeting, San Diego, CA, October.

Hall, P.M., Panousis, N.T., and Manzel, P.R. (1975). "Strength of Gold Plated Copper Leads on Thin Film Circuits Under Accelerated Aging," *IEEE Trans. on Parts, Hybrids, and Packaging*, PHP-11, No.3, pp. 202-205.

Harman, G.G. (1989). "Reliability and Yield Problems of Wire Bonding in Microelectronics," *A Technical Monograph of the ISHM*.

Harper, C.A. (1991). *Electronic Packaging and Interconnection Handbook*, McGraw-Hill, Inc., NY.

Hermansky, V. (1972). "Degradation of Thin Film Silver-Aluminum Contacts," *Fifth Czech. Conference On Electronics and Physics*, Czechoslovakia, October 16-19, pp. II.C-11.

Hirota, J., Machida, K., Okuda, T., Shimotomai, M., and Kawanaka, R. (1985). "The Development of Copper Wire Bonding for Plastic Molded Semiconductor Packages," *35th Electronic Component Conference Proceedings*, Washington, DC, pp. 116-121.

James, K. (1977). "Reliability Study of Wirebonds to Silver Plated Surfaces," *IEEE Transactions on Parts, Hybrids, and Packaging*, PHP-13, pp. 419-425.

Jellison, J.L. (1975). "Susceptibility of Microwelds in Hybrid Microcircuits to Corrosion Degradation," *13th Annual Proceedings of the Reliability Physics Symposium*, Las Vegas, Nevada, pp. 70-79.

Jones, R.M. (1975). *Mechanics of Composite Materials*, Chapter 2, Hemisphere Publishing Co., NY.

Kamigo, A. and Igarashi, H. (1985). "Silver Wire Ball Bonding and Its Ball/Pad Interface Characteristics," *35th Proceedings of the IEEE Electronic Components Conference*, Washington DC, 91-97.

Kumar, A.H. and R.R.Tummala (1991). "State-of-the-art, Glass-ceramic/copper, Multilayer Substrate for High Performance Computers," *The International Journal for Hybrid Microelectronics*, 14, 4, December.

Lampack, R.H. (1989). "Urethane Coatings," *Electronic Materials Handbook*, v. 1, Ed. M.L. Minges, ASM International, Materials Park, OH, 775-781.

Lang, B. and Pinamenni, S. (1988). "Thermosonic Gold Wire Bonding to Precious-Metal-Free Copper Leadframes," *38th Proceedings of the IEEE Electronic Components Conference*, Los Angeles, CA, pp. 546- 551.

Lau, J.H., Erasmus, S.J., and Rice, D.W. (1989). "Overview of Tape Automated Bonding Technology", *Electronic Materials Handbook*, v. 1, ed. M.L. Minges, ASM International, Materials Park, OH.

Levine, L. and Shaeffer, M. (1986). "Copper ball Bonding," *Semiconductor International*, pp. 126-129,

Libove, C., Perkins, R. W., and Kokini, K. (1982). "Microcircuit Package Stress Analysis," *RADC-TR-81-382*, Rome Air Development Center, NY.

Licari, J.J. and L.R. Enlow (1988). *Hybrid Microcircuit Technology Handbook*, Noyes Publications, NJ.

Lide, D.R. (1990). *Handbook of Chemistry and Physics*, 7th ed., CRC Press.

Lubin, G. (1986). *Handbook of Composites* Van Nostrand Reinhold Company, NY.

McCluskey, F. P. and M. Pecht (1997). "Pushing the Limit: The Rise of High Temperature Electronics," *Advanced Packaging*, Jan/Feb 1997.

McCluskey, F.P., Grzybowski, R.R., and Podlesak, T. (1996). *High Temperature Electronics*, CRC Press, Boca Raton, FL.

McCluskey, F. P., Condra, L., Fink, J., and Torri, T. (1996). "Packaging Reliability for High Temperature Electronics: A Materials Focus," *Microelectronics International*, 41, Sept. pp. 23-26.

McCluskey, F. P.,. Das, D., Jordan, J., Condra, L., Torri, T. and Fink, J., (1996). "Packaging of Electronics for High Temperature Environments," *Proceedings of the International Electronics Packaging Society Conference*, Austin, TX, pp. 143-151

Minges, M.L., (1989), *Electronic Materials Handbook*, v. 1, ASM International., Materials Park, Ohio,

Morio G., Ikeguchi, N., and Ayano, S. (1985). "BT Resin Widens the Choice of Laminate Materials," *Electronic Packaging & Production*, December, pp. 30-33.

Mumby, S.J. (1989a). "An Overview of Laminate Materials with Enhanced Dielectric Properties," *Journal of Electronic Materials*, 18, No. 2, pp. 241-250.

Mumby, S.J. and Yuan, J. (1989b). "Dielectric Properties of FR-4 Laminates as a Function of Thickness and the Electrical Frequency of the Measurement," *Journal of Electronic Materials*, v. 18, No. 2, pp. 287- 292.

Mumby, S.J. and Schwarzkopf, D.A. (1989c). *Electronic Materials Handbook*, v. 1, *Packaging*, Ed. Menges, M.L., ASM International, Materials Park, OH,

Neugebauer, C., Carlson, R.O., Filion, R.A. and Haller, T.R. (1991). "Multichip Module Design for High Performance Applications," *Multichip Modules*, IEEE Press, Piscataway, NJ, pp. 46-52.

Olsen, D.R. and James, K.L. (1984). "Evaluation of the Potential Reliability Effects of Ambient Atmosphere on Aluminum-Copper Bonding in Semiconductor Products," *IEEE Transactions on Components, Hybrids, and Manufacturing Technology*, CHMT-7, pp. 357-362.

Onuki, J., Koizumi, M., and Araki, I. (1987). "Investigation on Reliability of Copper Ball Bonds to Aluminum Electrodes," *IEEE Transactions on Components, Hybrids, and Manufacturing Technology*, CHMT-10, pp.550-555.

Oswald, R.G. and Miranda, R. (1977). "Application of Tape Automated Bonding Technology to Hybrid Microcircuits," *Solid State Physics*, March, pp. 33-38.

Pecht, M. (1991). *Handbook of Electronic Package Design*, Marcel Dekker, Inc., NY.

Pitt, V.A. and Needes, C.R.S. (1982). "Thermosonic Gold Wire Bonding to Copper Conductors," *IEEE Transactions on Components, Hybrids, and Manufacturing Technology*, CHMT-5, pp. 435-440.

Riches, S.T. and Stockham, N.R. (1987). "Ultrasonic Ball/Wedge Bonding of Fine Cu Wire," *Proceedings of the 6th European Microelectronics Conference (ISHM)*, Bournemouth, England, pp. 27-33,

Seraphim, D. P., Lasky, R. C., and Li, Che-Yu (1989). *Principles of Electronic Packaging*, McGraw-Hill Book Company, NY.

Shepler, T.H., and Casson, K.L. (1989). "Flexible Printed Boards," *Electronic Materials Handbook*, v. 1, Ed. M.L. Minges, ASM International, Materials Park, OH, 578-596.

Shigley, J.E. (1986). *Mechanical Design*, McGraw-Hill Book Company, NY.

Shugg, W.T. (1986). *Handbook of Electrical and Electronic Insulating Materials*, Van Nostrand Reinhold Company, NY.

Shukla, R.K. and Mecinger, N.P. (1985). "A Critical Review of VLSI Die-Attachment in High Reliability Applications," *Solid State Technology*, 28(67), July.

Speerschneider, C. and Lee, J. (1989). "Solder Bump Reflow Tape Automated Bonding," *Microelectronic Packaging Technology: Materials and Processes*, ASM, pp. 7-12.

Suhir, E. and Poborets, B. (1990). "Solder-Glass Attachment in CERDIP/CERQUAD Packages: Thermally Induced Stresses and Mechanical Reliability." *Transactions of the ASME Journal of Electronic Packaging*, 112, pp. 204-208.

Thomas, R.E., Winchell, V., James, K., and Scharr, T. (1977). "Plastic Outgassing Induced Wirebond Failure," *27th Annual Proceedings of the IEEE Electronics Components Conference*, Arlington, Virginia, pp. 182-187.

Trace Laboratories (1986). *Test Report on MLBs Fabricated from Du Pont's Non-woven Corlam 2643*, 8 May.

White, D. (1990). "New High Ground in Hybrid Packaging," *Report by Carbide Electronic Components*, Newark, NJ.

Zaky, A.A (1970). *Dielectric Solids*, Routledge and K. Paul, London, U.K.

INDEX